Selling Outer Space

STUDIES IN RHETORIC AND COMMUNICATION
General Editors:
E. Culpepper Clark
Raymie E. McKerrow
David Zarefsky

James L. Kauffman

Selling Outer Space

Kennedy, the Media, and Funding for
Project Apollo, 1961–1963

The University of Alabama Press   Tuscaloosa

Library of Congress Cataloging-in-Publication Data

Kauffman, James Lee, 1958–
    Selling outer space : Kennedy, the media, and
funding for Project Apollo, 1961-1963 / James L.
Kauffman.
        p.   cm.— (Studies in rhetoric and communication)
    Revision of author's thesis (Ph. D.)—Indiana Univeristy. 1989,
originally presented under the title: Selling space.
    Includes bibliographical references and index.
    ISBN 0-8173-0747-8 (cloth : alk. paper)
    ISBN 978-0-8173-5590-6 (pbk. : alk. paper)
    1. Project Apollo (U.S.)   2. Communication in politics—United
States.   3. United States—Politics and government—1961–1963.
4. Mass media—United States—Influence.   I. Title.   II. Series.
TL789.8.U6A5428   1994
387.8´0973—dc20   94-4653

To Joseph, Neal Thomas, and Kathi

Contents

Acknowledgments

Many people assisted in the preparation of this book, most notably my teachers at Indiana University. Ronald Lee, William Wiethoff, and James Andrews offered excellent criticism and advice as members of my dissertation committee. I owe a special debt to J. Michael Hogan, who directed my dissertation, and Robert Gunderson, who taught me how to write and encouraged my scholarship. I thank Edwin Rowley of Indiana State University, Steven Pullum of the University of North Carolina, Wilmington, and Harold Barrett (emeritus), John Hammerback, and Jack Samosky of California State University, Hayward, for their encouragement and sound advice. I also thank Marcia Brubeck for her thorough and precise editing and Nicole Mitchell, of the University of Alabama Press, for her assistance and prompt response to questions and requests.

For help with research, I am indebted to NASA archivist Lee Saegesser and to librarian Frances Wilhoit at Indiana University for their valuable assistance. I thank the Graduate School at Indiana University for funds that allowed me to visit the NASA History Office in Washington, D.C., and Indiana University Southeast for a summer fellowship that allowed me to do research on photography.

Part of this book originally appeared as "NASA's PR Campaign on Behalf of Manned Space Flight, 1961–63," in *Public Relations Review*, edited by Ray Hiebert, Spring 1991, pp. 61–66, copyright © 1991 by JAI Press Inc., and is used by permission.

On a personal note, I wish to thank my parents, Neal and Betha Kauffman, and my in-laws, Eugene and Elaine Dougherty, for their love and support.

Finally, a wonderful human being, Juli Crecelius, typed the manuscript.

Selling Outer Space

Introduction

The explosion of space shuttle *Challenger* on 28 January 1986 prompted serious questions about American space policy, particularly its emphasis on staffed rather than unstaffed flight.[1] Americans attempting to understand the development of United States space policy in the aftermath of the explosion might find it difficult to make a valid assessment. The thirty-year history of American space exploration has stood as a tribute more to human beings than to technology. Since the inception of America's space program, human beings have remained its focal point. Only in the aftermath of space disasters like the Apollo fire or the *Challenger* disaster have Americans seriously questioned the primacy—or even the need—of human beings in the exploration of space. Moreover, many Americans' interpretation of the early space program reflects Tom Wolfe's best-selling novel, *The Right Stuff,* and the movie based on that book. Knowing little of the actual events surrounding Project Mercury, Americans have had their perceptions colored by a fictionalization of the period. Finally, the program has become identified with astronaut-turned-senator John H. Glenn, Jr., and with perhaps the most romanticized president in American history, John F. Kennedy. Thus the crucial period in the creation of American space policy is steeped in myth. To understand the myths is to gain a better understanding of the origins and evolution of American space policy.

At the time, as historian Walter A. McDougall observes, the

Apollo moon program was "the greatest open-ended peacetime commitment by Congress in history."[2] Although NASA began in 1958, Project Apollo became the true beginning of the American space effort; Ralph G. Martin argues that the ten minutes Kennedy devoted to the lunar landing in his 25 May 1961 address marked the "real birth" of America's space program.[3] The budgets for manned space programs rose dramatically in fiscal years 1962 and 1963, as did NASA's overall budgets. Of NASA's $1.7 billion budget for 1962, $500 million went toward manned flight. The budget for fiscal year 1963 mushroomed to $3.7 billion, with $2.2 billion devoted exclusively to manned flight. And just one year later, the overall budget reached a whopping $5.1 billion—an increase of 300 percent in two years.

This book examines the Kennedy administration's rhetorical campaign to persuade Congress and the public to adopt a manned flight to the moon. In so doing, the study addresses three key themes. First, it illuminates the contrasting nature of technical and narrative arguments and explores how those arguments play different roles in public discussion of social policy. Second, the book examines how both the executive branch and the news media function to help set the agenda in American politics. In doing so, it offers a case study of the increasingly complex relationship between government and the media. Finally, it explores the power of technology to shape and direct human action. It begins by identifying the crucial events that led to the decision to place a man on the moon.

The year 1961 stands out as critical in the development of American space policy. One can trace America's preference for putting human beings into space back to the change in administrations. The National Aeronautics and Space Administration (NASA) and its first manned space program, Project Mercury, began under President Dwight D. Eisenhower's administration. Yet President Eisenhower never approved any of NASA's plans for a manned, lunar expedition or any steps leading toward that objective.[4] He virtually eliminated all spending for the proposed Gemini and Apollo projects, NASA's manned space missions to succeed Project Mercury.[5] Ike found no military or scientific justification for a lunar landing or for continued manned flight beyond Project Mercury. He refused to enter into a race to the moon with the Soviet Union.

John F. Kennedy, as John M. Logsdon writes, may well have won the 1960 election "partly because of space-related issues."[6] In his Pulitzer-Prize-winning history of the space age, Walter McDougall points out that no campaign issue "better symbolized" Kennedy's "New Frontier,"[7] the theme of Kennedy's presidential campaign. He stated this theme clearly in his nomination acceptance speech:

"We stand today on the edge of a new frontier—the frontier of the 1960s—a frontier of unknown opportunities and perils—a frontier of unfulfilled hopes and threats."[8] After only four months in office, Kennedy reversed Eisenhower's position, setting a new American space policy that would continue to the present. On 25 May 1961, in a speech to a joint session of Congress, Kennedy proposed that the United States land a man on the moon within the decade and asked Congress for supplemental appropriations for manned flight totaling over $500 million. With little debate, Congress granted Kennedy's request; the race to the moon had begun.

Of all the major problems Kennedy faced when he assumed the presidency, "he probably knew and understood least about space."[9] McDougall has written that in Kennedy's first few months in office, Kennedy hesitated to make "bold forays into this new frontier."[10] The press detected his reluctance. During a news conference, a reporter observed that the president did not "seem to be pushing the space program nearly as energetically" as he had during the campaign.[11] Two events in early 1961, however, apparently inspired Kennedy to act.

On 12 April 1961, Soviet cosmonaut Yuri L. Gagarin shocked the nation by becoming the first man to orbit the earth. Kennedy immediately asked Vice President Lyndon B. Johnson to evaluate America's space program and determine whether the United States stood a chance of besting the Soviets in space. Within a week of Gagarin's flight, the nation also learned of the Bay of Pigs fiasco. In his despair, Kennedy began to view space achievement as a way to erase the memory of the Bay of Pigs and as "a symbolic manifestation of national determination and vitality."[12] Before he would act, he needed a sign that would convince him, and the nation, that the U.S. space program could compete with the Soviets'. With the success of Alan B. Shepard's suborbital flight on 5 May 1961, Kennedy became confident enough to propose the goal of landing a man on the moon.

Kennedy viewed the American space program primarily in "political terms."[13] He knew that reaction to Gagarin's flight presented him with an opportunity to "identify himself with a new space program with wide public appeal."[14] He had also consulted with enough members of Congress to realize that he would probably gain "substantial political benefits" by advocating a vigorous space program.[15] McDougall contends that the decision to go to the moon went well beyond improving Kennedy's influence in Congress and his standing in public opinion. Project Apollo reflected a fundamental change in the philosophy of government promoted from the White House.

Eisenhower had been genuinely conservative in his views of the presidency. He believed in restricting the federal government to its traditional roles and strictly maintaining the separation of the public and private sectors. The new administration, by contrast, came to view Project Apollo as part of a broader challenge to the traditional role of the federal government. Proponents of the Apollo program within the administration were dissatisfied with the existing management of national resources. Viewing the space program as a "catalyst" for social progress, technological revolutions, and the restructuring of institutions in ways dimly foreseen but assumed to be progressive, this new group brought to office the belief that the cold war would be won or lost in the Third World. Prestige, they contended, would be as important as power in the struggle. This belief justified an increased role for government in the management of the national image. With pressures to demonstrate leadership in light of the Bay of Pigs setback, the administration viewed an active space program as the first step in regaining some of America's lost credibility.[16]

Not only did the commitment to a lunar landing signal a more active presidency and a greater concern with national image under Kennedy, but it also promoted what McDougall calls technocracy, or the "institutionalization of technological change for state purposes."[17] This explosion of research and development, funded by the state, in turn implied greater centralization of government. Advanced technology would require mobilizing science, education, and government. Only by pulling together these interests could technocracy succeed. If such a program were adopted, however, it would redefine the relationship between the public and private spheres; the two would become more interdependent.

In the fervor of the cold war and the space race in the early 1960s, the technocratic faith went unchallenged, as evidenced by the tremendous increases in federal spending for education, science, and research and development. Congress passively accepted the new technocracy, raising few questions or objections. Although Americans would sometimes dispute the goals their government ought to pursue, few challenged the faith in technology. Traditional views concerning the need for balanced budgets and limited government gave way to the technocratic temptation.

The Kennedy administration justified Project Apollo with both technical and narrative arguments. Technical argument deals with technical or scientific issues and requires "specialized forms of reasoning" where "limited rules of evidence, presentation, and judgment are stipulated in order to identify arguers of the field and facilitate the pursuits of their interests."[18] In short, technical argument is the argument of experts. Narrative argument, in contrast, assumes a different kind of rationality and is based on storytelling.

Walter Fisher characterizes the different kinds of arguments above in terms of a "rational world paradigm" and a "narrative paradigm" of human communication.[19] According to Fisher, the rational world paradigm is based on "subject matter knowledge, argumentative ability, and skill in employing the rules of advocacy in given fields."[20] The traditional paradigm assumes that the experts know more than the ordinary citizen. In the narrative paradigm, rationality is determined by "the nature of the persons as narrative beings—their inherent awareness of narrative probability, what constitutes a coherent story, and their constant habit of testing narrative fidelity, whether the stories they experience ring true with the stories they know to be true in their lives."[21]

The Kennedy administration offered political, scientific, military, and economic justifications for sending a man to the moon. It attempted to appeal to the broadest possible audience. As justified by the administration's experts, the program had something for everybody. Moreover, the administration depicted Project Apollo as a structurally coherent narrative. Ultimately, the narrative constructed by the Kennedy administration and reinforced by the media produced a compelling case for funding manned space flight that could not easily be refuted by technical claims grounded in a more "rationalistic" worldview.

Kennedy effectively integrated his space policies with the New Frontier theme. The New Frontier included many of the constituents of the old frontier: adventurous and independent pioneers willing to battle evil enemies and tame a hostile, unknown environment. Kennedy chose to define the space program in ways that would symbolize the optimism and excitement of traditional frontier stories. Calling the world's "battle" between "freedom and tyranny" the "battle for men's minds," Kennedy proposed that no "adventure" had greater impact on men's minds than "the dramatic achievements in space."[22] Space as a New Frontier, however, presented a problem; it was infinite and intangible. How could one "explore" or "dominate" outer space? Kennedy solved this problem by choosing a concrete goal: landing a man on the moon. "Until we have a man on the moon," he proclaimed, "none of us will be satisfied."[23] The romance and intrigue of the moon, coupled with the story of the frontier, gave Kennedy a way of depicting a march to the moon that was both exciting and familiar to Americans. It used a narrative form—a heroic adventure story—complete with heroes and villains.

The chief hero was, of course, John Glenn. On 20 February 1962, Glenn blasted into space, becoming the first American to circle the earth. Although America launched longer, more difficult flights in the succeeding months, none compared in popularity with Glenn's. His triple orbit captured the imagination of the American people. He became an instant hero. McDougall terms America's outpouring of emotion for Glenn's feat "a national catharsis unparalleled in the quarter century of the Space Age."[24] Yet American enthusiasm went well beyond admiration for the flight itself. America might have celebrated anyone who put it back in the space race. But clearly Glenn was special. Americans "found Glenn the man fully the equal of Glenn the astronaut."[25] Much was made of Glenn's personal qualities. According to *Time*, "Glenn's modesty, his cool performance, his dignity, his witticisms, his simplicity—all caught the national imagination."[26]

Although scholars have devoted a great deal of time and energy to studying the space program, few writers have addressed the Kennedy administration's rhetoric promoting it. Surprisingly, biographies and books on Kennedy's administration say little about Kennedy and the space program. When they do comment, they discuss his hesitation about committing to a space program at the start of his administration or how Gagarin's flight and the Bay of Pigs prompted him to take action. A few mention his major address on space delivered on 25 May 1961. Few, however, devote more than two or three pages to Kennedy's views or his relationship to the program.[27] Scholars have written about Kennedy's presidential rhetoric. But only one focuses on his rhetoric concerning the space program.[28] And that study examines depictions of public time by Kennedy and Richard Nixon in the space debate during the 1960 presidential campaign.

One would expect scholars to have examined more exhaustively the rhetoric of one of America's most noted presidential communicators. This is not the case. One writer has explored the notion of space as the "New Frontier." Janice Hocker Rushing's work examines the changing notion of the frontier, from Frederick Jackson Turner's frontier thesis to Kennedy's New Frontier. She presents a provocative and insightful analysis of how locating the New Frontier in space alters the frontier myth and its explorers. The work helps explain why Glenn, of all the astronauts, gained such tremendous fame. Her articles, however, say little about Kennedy's rhetoric. Rushing analyzes popular movies about outer space and the frontier rather than the actual political rhetoric of early 1960s. Most important, she says nothing about the possible influence of the New Frontier myth on political decisions affecting the space program, particularly in Congress.[29]

Historians have written more extensively about the politics of the space program, but they have said little about the rhetoric of the times. Loyd S. Swenson, Jr., James M. Grimwood, and Charles C. Alexander provide a detailed history of Project Mercury, NASA's first manned space program. The study covers Kennedy's presidency and the events surrounding the *Mercury* flights and astronauts.[30] At least three historians have examined the rise of Project Apollo from 1961 to 1963. Vernon Van Dyke examines congressional debate over Apollo, chronicling the motivations for a moon landing. Yet his book says nothing about how the Kennedy administration argued for the program.[31] Logsdon's study of Project Apollo focuses on Kennedy's decision-making process. Finally, McDougall looks at the political motivations of the principal participants in the space program and provides a keen analysis of Kennedy's reasons for deciding to commit the nation to a moon landing. None of these historians provides an in-depth analysis of the administration's arguments in favor of the space program or it's rhetoric as it may have influenced congressional debate over funding.

A few scholars have commented on the media's coverage of the space program. Many of the studies attack the media's coverage as uncritical, biased, and manipulative. Robert Cirino criticizes the media for failing to raise more questions about Kennedy's proposal to go to the moon.[32] Edwin Diamond, who covered the space program for *Newsweek*, calls the media's coverage manipulative.[33] Wolfe's famous fictionalized account of the space program criticizes the media for uncritically accepting, and even adding to, the romanticized rhetoric of the administration.[34] And Frank Van Riper, author of Glenn's biography, says some of *Life* magazine's coverage would be better viewed as NASA public relations.[35] Not all researchers, however, view the press as biased. Ronald E. Ostman and William A. Babcock, responding to Cirino's work, examine three newspapers' coverage of Kennedy's press conferences and conclude that the press was not biased toward Kennedy's position on the space program.[36] Thus controversy persists over the media's role in gaining support for Project Apollo.

Existing literature leaves unanswered many important questions about the relationship between the rhetoric of the Kennedy administration, media coverage of the space program, and deliberations within Congress over funding for space. Excluding Kennedy's campaign speeches and his inaugural address, Theodore O. Windt, Jr., notes, communication scholars have written "hardly anything" about "Kennedy's rhetoric." He labels the lack of study "astonishing." Although Kennedy's rhetoric established the "themes used or abused by subsequent administrations, it has been subjected to precious little scholarly inquiry."[37] Similarly, there has been little close analysis of the media's coverage of the space program or congressional deliberations. The present book aims to fill these gaps.

Richard E. Neustadt, in his classic study of the presidency, observes that "presidential power is the power to persuade."[38] This statement, as Windt notes, clearly located "the scholarly place of rhetoric within presidential studies."[39] Interest in the rhetoric of the presidency has exploded in the last decade, as more and more scholars insist that the presidency "is based on words, not power."[40]

Much has been said about the president's power to set the national agenda.[41] But the president's power to persuade goes well beyond his agenda-setting functions, for the president also shapes the ways people interpret or give meaning to events through the words he uses to describe those events. "Language," writes Murray Edelman, "evokes most of the political 'realities' people experience."[42] Language does not merely reflect political realities; it plays an active role in *creating* them by "organizing meaningful perceptions abstracted from a complex, bewildering world."[43] Few political events may be observed directly, and the language used to describe those events "shapes the meaning of what the general public and government officials see."[44] In short, words may largely determine the political world one sees, and the president's words carry special weight in national politics.

One way the president's language may influence perception is through definition. The power to persuade, writes David Zarefsky, is largely "the power to define."[45] By naming an object or an idea, one influences attitudes about it. One might argue that choosing a definition is tantamount to pleading a case. "The president's goal," Zarefsky adds, "is to place ambiguous situations into a context such that the presumed response is congenial to his purpose."[46] Choosing among the numerous symbols available to characterize a situation, the president also hopes to influence the symbolic choices of others. President Johnson's characterization of the attempt to eradicate poverty as a "war," for example, not only aimed to influence the way Americans viewed the administration's policies but also aimed to influence the types of symbols, or characterizations, that other political actors would choose.

The president is not alone in his attempt to gain support for a program or a policy. In debate over the space program, President Kennedy's allies included a number of advocates within the administration whose power and influence, and whose very existence in some cases, depended upon gaining support for the president's space program. Chief among these allies were Vice President Lyndon B.

Johnson and NASA. My book shows that the president has the power not only to persuade Congress and the public personally but also to unleash powerful bureaucratic agencies that can advocate on behalf of specific policies or priorities.

To say that the president has considerable power to set the agenda and to define the terms of national debate, or to say that he often commands a virtual army of political advocates, is not to say, however, that the president is all powerful in political debate. In recent history, the media's challenge to official portrayals of the war in Vietnam demonstrated that presidents do not have the power to dictate the news that Americans receive. Yet Robert Cirino characterizes the media coverage of Project Apollo as an "electric spectacle and advertising campaign calculated to sell the moon landing to the American people."[47] If so, Kennedy may have been unusually successful in defining the program in ways that influenced the media's coverage. Cirino's statement, and the tremendous amount of coverage given to the early space program, warrant a more detailed analysis of the relationship between the Kennedy administration's depiction of the American space program and coverage of the story by the major news media. It appears that the Kennedy administration viewed the media not as a neutral conduit of information to the public but as a valuable ally in establishing a domestic policy priority.

Finally, the space program illustrates an increasingly problematic relationship between technological systems and human choice. Financially, technically, and scientifically, unstaffed space exploration appears superior to staffed exploration in many ways. In developing a space program that privileges human beings, NASA created a dilemma for itself. To maintain popular, political, and financial support for the space program, NASA must continue to emphasize the frontier mythology. In so doing, however, NASA may in the end direct American space policy toward manned space programs that sustain the myth but are technically and scientifically inferior to unstaffed programs of competing nations. Thus, my book indicates an irony of technological advance: the more inventions human beings create to control their environment, the more the inventions shape the environment and thereby "control" or limit human choice and action.

An administration's power to persuade is often sorely tested in the halls of Congress. The administration's success in defining the space program in a fashion conducive to its goals was ultimately measured in Congress. Was the Kennedy administration's portrayal of the space program echoed in Congress? How did the administration create a climate of acceptance of the space program? This book ultimately seeks to answer these questions by examining the perceived realities reflected and promoted in the debate over a manned lunar landing from 1961 to 1963.

I begin by exploring the Kennedy administration's attempts to sell Project Apollo to Congress and the nation. The first two chapters examine the two very different ways in which the administration sought to peddle the program. Chapter 1 describes NASA's public relations campaign on behalf of a manned lunar landing and reviews the major political, scientific, military, and economic arguments employed. Chapter 2 discusses the administration's efforts to depict the manned lunar landing in narrative form as a great frontier adventure story complete with heroes and villains.

The book's third and fourth chapters assess the coverage of the manned moon mission in the popular print media and argue that the media largely echoed the administration's rhetoric instead of offering critical, objective reportage. Chapter 3 identifies journalistic values that account for both positive and negative coverage of the program but that ultimately supported—and at times even glorified—manned exploration. Chapter 4 uncovers the most striking example of the media's celebration of the space program: *Life* magazine's coverage. Going well beyond mere cheerleading, the magazine established a legal and financial relationship with the astronauts, making *Life* little more than an extension of NASA's public relations program.

The last chapters analyze the argumentative substance of the deliberations over Project Apollo in congressional space committees and on the floor of Congress. Chapter 5 explores the deliberations of congressional space committees and posits that committee members never seriously challenged the need for or feasibility of a manned lunar landing. Instead, the congressional space committees acted as advocates who sought ways to justify the program to the public and their fellow members of Congress. Chapter 6 examines debates on the floor of Congress over funding for Project Apollo. It explains that although space critics offered serious and persuasive attacks on the political, scientific, military, and economic justifications for a manned moon mission, Congress still approved enormous sums for the program. The only strong rationale for Project Apollo that went unrefuted was its depiction as a frontier narrative. The chapter concludes that critics failed to counter the narrative for one of three reasons: they found the story persuasive, they offered inadequate refutation, or they did not wish to suffer the adverse political consequences of maligning such a nationalistic, patriotic story or its heroes.

America's manned space program in the early 1960s constituted perhaps the greatest commitment of national resources to a peaceful pursuit in the history of the nation. The Kennedy administration's campaign was so successful that in the succeeding thirty years, Americans have rarely questioned either the cost or the need for a large, staffed space project. By examining the ways in which the Kennedy administration persuaded Americans to adopt such a program, this book aims to gain insight not only into the persuasive efforts of the Kennedy administration but also into the processes by which American domestic policy issues generally are created, debated, and resolved.

1

The Kennedy Administration's Lunar Campaign

Four days before President Kennedy asked Congress to commit the nation to a manned lunar landing, Hugh Dryden testified before the Senate Appropriations Committee. Asked what practical use he saw in putting a man on the moon, NASA's second in charge admitted: "It certainly does not make any sense to me."[1] Dryden's disclosure seems remarkable in light of his position, the committee he addressed, and the remarks of the president just four days later. One might argue, however, that Dryden reflected the state of NASA's public position at the time. The president had not yet committed the country to a manned moon mission. After the president's speech, NASA officials testifying before congressional committees never failed to offer justifications for a manned lunar landing. One might explain Dryden's statement by examining Kennedy's initial reluctance to act on space matters.

Consumed with foreign policy matters during his first months in office, President Kennedy devoted little time to the space effort. He did, however, appoint a new NASA administrator, James Webb. Supported by his chief lieutenants, Webb immediately tried to fix some of the damage done by President Dwight D. Eisenhower's call for the elimination of manned flight programs beyond Project Mercury. Webb focused on two goals: refuting scientific criticisms aimed at Mercury and correcting the agency's budgetary deficiencies. The newly appointed Webb, lacking a clearly stated space objective, did his best to appease an anxious Congress as he waited for Kennedy to act.

On May 25, 1961, President Kennedy delivered what many termed his "second inaugural." The latter part of the speech was the most significant. Kennedy stated that to win the current "battle" around the world between "freedom and tyranny," America must engage in a great new "enterprise." The president proposed that within the decade, America should commit itself to landing a man on the moon and returning him safely to earth. Specifically, Kennedy asked Congress to approve an additional $531 million for NASA in fiscal 1962 and to make a five-year commitment of between $7 billion and $9 billion for a manned lunar landing. He disliked asking Congress for new appropriations, he concluded, but this is "a most serious time in the life of our country and in the life of freedom around the world."[2]

Kennedy's speech dramatically elevated space to the top of his agenda. In establishing a specific goal and calling for a specific program, Kennedy committed the entire nation to, and created a political climate for, manned space exploration. In short, Kennedy's speech launched the campaign to sell Project Apollo to Congress and the nation.

NASA's Public Relations Campaign

The Kennedy administration, writes Vernon Van Dyke, hoped to influence attitudes "not simply through space shots but also through publicity concerning them."[3] One typically thinks of NASA as an agency concerned with science and technology, not with public image making. Yet in the early 1960s, the agency labored long and hard to create a positive image of its programs, its flights, and its astronauts. All along, however, it denied doing so. In short, part of the perception of NASA as a scientific, technological agency that was above manipulating public perception comes from the agency itself. Without doubt, NASA engaged in a sophisticated public relations campaign in trying to peddle Project Apollo.

NASA developed a large, active public relations department to support the agency's pitch. The space agency developed a general public affairs office (PAO), with smaller PAOs at each space center that were responsible for the public relations activities of that center. Within the center's PAO resides a public information office (PIO), designed primarily to handle the media. The leadership of the agency's PAO changed numerous times from 1961 to 1963. The aim of the PAO remained the same: to popularize and to help sell NASA's programs.

In 1959, NASA appointed an experienced public relations officer from the military to serve as the astronauts' liaison with the public and the press. Lieutenant Colonel John "Shorty" Powers became known as the "voice" of the astronauts. He exercised strict control over the media's access to the astronauts and taught the Mercury team how to project the image NASA desired. At times he went so far as to put words in the mouths of the astronauts. Powers relayed to the media, for example, the astronauts' transmissions to ground control during their flights. During Shepard's flight, Powers reported that the astronaut said everything was "A-Ok." The public and the press became enamored of the expression. A later review of audiotapes of Shepard's transmission demonstrated that he never used the expression. Powers had invented it.[4]

Besides working with the astronauts, NASA took great pains to prepare its officials for questions about its programs, especially those officials who spoke before congressional appropriations and authorization committees. For example, a memorandum from the public information office to James Webb on 21 May 1961 provided pages of suggested answers to anticipated questions from the news media about the manned space program. NASA not only coached officials on how to testify at congressional hearings but also provided material describing and justifying its programs directly to the committees. The volume of material grew steadily, so that by 1963, Webb submitted to members of congressional space committees four volumes of "justification books" containing a "full breakdown of each and every part of the program" and explaining "what, where, how, and when work will be done and what it will cost."[5] NASA officials used the material frequently during congressional hearings. NASA also wished to provide a consistent, persuasive message to the public. In mid-1963, NASA began sending its public information officers a letter excerpting space quotations from NASA officials. A memo on the subject expressed its author's hope that the letter would provide answers to "questions which NASA employees might meet either in public speaking appearances or in other contacts with the general public."[6]

With increased criticism in 1963, NASA administrators dramatically increased their speaking appearances to sell the program. In 1962, for instance, Webb delivered forty-nine public speeches and Dryden sixteen. In just the first six months of 1963, Webb had already presented forty-two addresses and Dryden twenty-one.[7] Obviously, NASA took great care in presenting a consistent, coherent image of the agency and Project Apollo.

NASA developed a public relations strategy aimed at gaining positive exposure for its personnel, organization, and activities. Most obviously, the agency used President Kennedy to call attention to the program. When NASA needed publicity for its flights, especially around the time of budget hearings, it had the president present the returning astronaut with NASA's own Distinguished Service Medal in a ceremony in the White House Rose Garden. One might argue, moreover, that the award itself was purposefully designed as a public relations vehicle, created to allow NASA to call attention to its officials and the astronauts. Interestingly, Kennedy made the presentation only after the flights of Alan Shepard, John Glenn, and Gordon Cooper. NASA administrator James Webb presented the award to the other astronauts. As expected, the president's presentations received widespread coverage in the press, which drew attention to the astronauts and their successful flights. Similarly, when NASA came under attack by critics in late 1962 and 1963, the president went on a tour of the space facilities in the South to call attention to the program. He used the opportunity to make his only speech dealing exclusively with space at Rice University on 12 September 1962. Aimed at blunting criticism of the program's mounting cost, Kennedy tried to rally support for a moon shot. When Congress seemed intent on slashing NASA's budget in late 1963, Kennedy again took to the road, partly to call attention to the program by touring Cape Canaveral and other southern facilities.[8] The trip took him to Dallas, where he was killed.

The space agency also staged or sponsored events to gain publicity. The day after the president made his speech committing the country to a manned moon mission, NASA attempted to maintain high visibility for the space program by cosponsoring the First Conference on the Peaceful Uses of Outer Space in Tulsa, Oklahoma. The city's chamber of commerce served as the other sponsor for the meeting, which gave administration officials like Vice President Lyndon Johnson and NASA administrator Webb the opportunity to make speeches about the program. In short, NASA staged the event to gain support for the president's proposal. The space agency continued with the conference in the succeeding years, holding it in Seattle in 1962 and in Chicago in 1963. NASA staged other events that served its purposes. When criticism of Project Apollo began surfacing in late 1962, the agency provided the fiscal sponsorship for a meeting of a group of scientists sympathetic to Project Apollo to study the problem of space exploration. Not surprisingly, the group produced a report strongly endorsing manned space exploration and countering the opinions of scientists critical of Project Apollo.[9] One finds in late November 1962 still another example of how NASA strategically used events to publicize itself and its activities. NASA, along with its cosponsor, the *Cleveland Plain Dealer*, staged the Space Science Fair. The ten-day exposition on America's conquest of space offered Cleveland residents free movies and speakers, as well as special programs for children and schools.[10] These examples demonstrate NASA's public relations strategy. When it came under fire, NASA created events that publicized itself and its activities and helped blunt criticism aimed at the space program.

The space agency made its greatest public relations effort in publicizing the program through the astronauts and their flights. Again, it managed these events in ways that garnered publicity and created a positive image of the astronauts, their flights, and the program. The agency never missed a chance to exploit public enthusiasm for the astronauts and their flights. John Glenn's flight serves as a good example. First, it seemed more than a coincidence that the flight took place exactly one week before the House Committee on Science and Astronautics opened hearings on NASA's fiscal year (FY) 1963 authorization bill. Second, besides receiving NASA's Distinguished Service Medal from the president at the White House, Glenn appeared in ticker-tape parades in Washington and New York and became one of a few people ever to address a joint session of Congress. In addition, flanked by Alan Shepard and Gus Grissom, Glenn appeared as the star witness on the opening day of hearings in the Committee on Science and Astronautics on NASA's authorization bill. The comments of chairman George Miller suggest that the timing of Glenn's flight was not coincidental. Representative J. Edgar Chenoweth (R., Colo.) commented to NASA administrator Webb that in light of Glenn's successful flight just one week earlier, the agency had picked "a most opportune propitious time" to present its budget request. Chairman Miller retorted, "May I say to the gentleman that that wasn't accidental."[11]

Scott Carpenter's flight casts more doubt on the notion that the timing was merely coincidental. NASA's FY 1963 authorization bill reached the House floor on May 23, 1962. Only hours later, on the morning of May 24, 1962, Scott Carpenter was scheduled to blast into space, becoming America's second man to orbit the earth. The House approved the bill unanimously, 343–0. Undoubtedly, one can observe NASA's public relations efforts most clearly with Gordon Cooper's flight in 1963. In a shrewd public relations move, NASA scheduled Cooper's mission on the anniversary of Charles Lindbergh's historic flight. Obviously, the agency wished to invite comparisons between the two men and their flights. Originally, the agency planned only two activities for Cooper after the flight: a presidential awards ceremony at the White House and a meeting with selected House leaders in the Speaker's office. When NASA saw the public reaction to the flight, it attempted to exploit the enthusiasm, adding a motorcade down Pennsylvania Avenue, a ticker-tape parade, and another speech to a joint session of Congress. Administration officials frankly acknowledged that they wanted to use the events to show Congress how much the public supported the program.[12]

The numerous "coincidences" lead one to question the part that NASA's public relations officials played in all of the astronauts' appearances. In late 1963, for example, the astronauts and their wives traveled to Washington to receive an award in a White House ceremony. Coincidentally, the ceremony took place on the exact day the House conducted crucial debates on NASA's appropriation bill. Conveniently, chairman George Miller (D., Calif.) invited the astronauts and their wives to visit the House chamber during the actual floor debates.

In short, the space agency staged numerous events to gain publicity and create a positive image of the astronauts, the agency, and its activities. The timing of many of the astronauts' flights and their appearances at important congressional debates on funding for NASA's programs seem more than coincidental. Although one can find no actual proof that NASA purposefully scheduled its flights to conform to congressional hearings and debates, the timing must certainly have played a part in the deliberations of members of Congress and in the enthusiasm of the public during crucial times in the congressional budgetary process.

NASA took no chances when it came to its image, especially with regard to Congress. It understood how negative congressional perceptions could hurt funding for its programs. Although public support was crucial for NASA's manned moon shot, Congress held the purse strings. James Webb had been director of the Bureau of the Budget from 1946 to 1949, and he demonstrated his sensitivity to the budgetary process in a memorandum of 30 November 1962 to President Kennedy. Responding to Kennedy's suggestion that a supplemental bill be submitted for FY 1963, Webb advised that a supplemental would "adversely affect" appropriations for FY 1964 by allowing critics to charge that "costs are skyrocketing" and that "'overruns stem from poor management'" rather than encouraging them to focus on NASA's progress. Undoubtedly the increase from $3.7 billion for 1963 to a potential $6.2 billion for 1964 would "raise more questions" than the previous year's jump, Webb felt. If your FY 1964 budget supports NASA's $6.2 billion request, Webb explained, NASA feels "reasonably confident" it can "work with" committees and congressional leaders to "secure their endorsement." I would prefer to make my "main fight" for the $6.2 billion request for 1964, he declared.[13] Webb's comments demonstrate a tremendous concern for NASA's image. He understood too well how Congress would perceive a supplemental bill. Once again, Webb's remarks demonstrate NASA's consciousness of public relations.

In staging events and making speeches, the Kennedy administration and the PAO also appeared very sensitive to gaining media coverage. President Kennedy and administrator Webb both denied charges that the administration had tried to manipulate press coverage of Alan Shepard's flight. Administration and NASA officials had merely responded to press and television requests, the two men argued. The administration took a number of steps to encourage press coverage of the flights. It sent invitations to members of Congress, it distributed press kits days in advance of the flights, it allowed television networks to set up on the Cape days in advance, it polled the press on its requirements weeks ahead of time, it made NASA officials available for interviews before the flights, and it provided reams of material on every conceivable aspect of the program.[14] Between 1961 and 1963, the space agency held numerous press conferences with the astronauts and NASA officials. It released hundreds of news releases and sent out elaborate, unsolicited press kits to the media.

NASA's most blatant step to ensure positive press coverage came in 1959 when the agency allowed the Mercury astronauts to sign a contract with *Life* for the exclusive rights to their personal stories. Despite criticism from other media, it renewed the contract in 1962. In addition, in early 1963, Wernher von Braun, director of Marshall Space Flight Center, accepted *Popular Science*'s invitation to write a monthly column on space. The administration, so successful in cultivating a positive image of the space program in 1961 and 1962, seemed to lose its Midas touch in 1963.

In the later part of 1963, the administration took two steps that turned into public relations crises and aided opponents of Project Apollo. First, in July, NASA offered the Columbia University School of Journalism a three-year $393,000 grant to study how to disseminate news about the space program and how to train science writers. The grant would bring students into NASA to do interim work producing scientific reports. The university, undoubtedly with prompting from the agency, called off the contract when critics began charging that the grant was aimed at improving NASA's public relations.[15] Two months later, President Kennedy surprised Washington by proposing a joint U.S.-U.S.S.R. moon expedition. Faced with growing pressure to reduce spending, the administration hoped Congress would view the proposal as a first step away from the costly commitment to land a man on the moon by 1970. The proposal, however, seemed to reverse the administration's position of beating the Soviets to the moon. Congressional space advocates were stunned. In one short speech, Kennedy undercut some of the most persuasive arguments for a moon shot: pride, prestige, and national defense. Congressional opponents of the space program took advantage of the confusion caused by the proposal to blast the moon shot once again.

Two other events beyond the administration's control also caused problems for NASA's image makers. During critical debates over NASA's budget in late 1963, both NASA and the General Accounting Office (GAO) released reports critical of NASA and the space program. The GAO report, ordered by the House Committee on Science and Astronautics, blamed NASA for waste and poor management. At almost the same time, NASA released its own report criticizing the quality of equipment provided by private contractors. Space advocates attacked the timing of the studies, which they argued provided ammunition for opponents of the program.

Representative George Miller, chairman of the House Committee on Science and Astronautics, attempted to defend the timing of the reports and to downplay their findings. Miller explained that prior to their release, the press had asked about the reports and that there was no valid reason for suppressing the documents. Miller quickly tried to minimize the importance of the documents. The GAO report, Miller argued, merely corroborated information published the previous year. He added that many of the shortcomings cited in the report had already been corrected. NASA also tried to temper its own criticism of poor workmanship by contractors at a two-day conference summarizing Project Mercury's achievements. The agency not only had its top officials appear at a news conference at the meeting, but it also took the unusual step of permitting a vice president from the McDonnell Aircraft Corporation, one of its prime contractors, to appear with its officials. The group attempted to deflect criticism of poor workmanship by stressing the overall success of Project Mercury.[16]

Despite the emergence of these difficulties in 1963, Congress continued to be generous in its support of NASA and approved $5.1 billion for the space agency. NASA's large professional public relations program had created a climate of support for Project Apollo that the new controversies did little to damage. NASA publicized the agency, the astronauts, and their flights and countered its critics by effectively shaping media coverage. Ultimately one must look to the substance of the case for a manned lunar mission if one is to understand the success of NASA's public relations. In the final analysis, the persuasiveness of the arguments for going to the moon, and not merely their timing or the media that carried them, accounts for congressional and public support for the manned lunar mission.

The Case for a Manned Lunar Landing

NASA's justification for a manned moon mission evolved from 1961 to 1963 into a coherent, sophisticated pitch that mustered a number of political, scientific, military, and economic arguments for going to the moon. Hugh Dryden's testimony before the Senate Appropriations Committee on 21 May 1961 demonstrates NASA's persuasive intent and its concern for the public perception of a manned moon mission. NASA had to do a great deal to help the public understand a manned lunar landing if the public was going to "accept it," Dryden explained. "And we are naturally concerned here with that." The program will "cost each American $100. Now whether Americans want to accept that for scientific reasons, prestige reasons, propaganda reasons, or all three, is something we up here must decide."[17] Dryden's statement not only sheds light on his own view of NASA's relationship with the public, but it also demonstrates Dryden's perception of the Senate committee as a partner in NASA's sales campaign.

The administration peddled a manned lunar landing on a variety of grounds. First, representatives of the administration argued that America had to embark on a manned lunar mission to protect its image abroad. Keenly aware of the propaganda value of space spectaculars, the president sent a memorandum to Vice President Johnson shortly after Gagarin's flight asking, "Is there any other space program which promises dramatic results in which we could win?"[18] Kennedy's remarks at a press conference the following day demonstrate the importance he placed on achieving a dramatic space shot. When questioned about America's future plans in manned space flight, Kennedy stated that the United States must determine whether any program, "regardless of cost," would enable it to be "first in any new area."[19] The image-conscious Kennedy restated his concern for the symbolic meaning of Soviet spectaculars in his 25 May address. The Soviets' "dramatic achievements" in space would have an impact on the minds of people throughout the world who were deciding whether to side with the United States or the Soviet Union. He returned to this theme often. Dramatic results in space meant nothing less than world leadership. "No nation which expects to be the leader of other nations," Kennedy remarked, "can expect to stay behind in this race for space."[20]

Kennedy's preoccupation with image is no surprise, considering the advice he received from his chief advisers, James Webb and Vice President Johnson. In response to Kennedy's memorandum asking whether the United States could do anything to attain a space spectacular, Johnson stated that "regardless of their appreciation of our idealistic values," other nations will "align themselves with the country they believe will be the world leader." Countries identify "dramatic" space accomplishments as "a major indicator of world leadership."[21] Webb offered similar advice. In a meeting over NASA's first budget supplement in 1961, Webb said that unless Kennedy reversed Eisenhower's elimination of manned space flight beyond Mercury, the Russians would "beat" America to "every spectacular exploratory flight" in the next five to ten years.[22] Although the United States had an advantage over the Soviet Union in overall space sciences, Webb shared the view of many who believed space leadership was symbolized by dramatic manned space flights. "The Soviets have demonstrated how effective space exploitation can be as a symbol of scientific progress and as an adjunct of foreign policy," he said. "We should not fail to recognize its potential."[23]

Webb, and other NASA officials, elaborated the image argument in a number of forums. In an address to a group of school administrators two days before Glenn's flight, Webb observed that other countries equated space successes with future world leadership. Unless we compete strongly with the Soviets in space, cautioned Webb, "our national prestige will suffer in the eyes of other nations." The United States should use space achievements as "a device of foreign policy and diplomacy," Webb advised, "to establish a definite position of leadership."[24] George Low, deputy director (programs), Office of Manned Space Flight, echoed Webb, calling American preeminence in space essential to "world leadership and prestige."[25] Perhaps NASA's deputy administrator, Hugh Dryden, summed it up best when he stated that merely "putting a man on the Moon" was not "very significant." The national effort to put him there, however, was indispensable to the country's image, he explained. "This is a symbol."[26]

Besides improving America's image, administration advocates argued that a manned lunar mission would advance science and technology. President Kennedy used this argument often, proposing that a moon shot would develop a "new frontier of science."[27] He characterized Shepard's flight as "an outstanding contribution" to the advancement "of space technology," and he praised Glenn's flight for "its scientific achievement."[28] Glenn seconded Kennedy's claim, saying that manned space flights lead to "across-the-board scientific advances."[29] One can find similar testimony from NASA officials.

In a memo from public affairs officer Bill Lloyd to James Webb, NASA anticipated two major questions about the scientific value of the manned lunar landing and suggested appropriate answers. "Does our scientific community support the effort to land a man on the moon?" the memorandum reads. The suggested answer: "I don't speak for the scientific community. However, I believe this program will prove to be a tremendous stimulus to all science. There is virtually no major area of science that will not feel this impact, and ultimate benefit." Similarly, the second question asks, "Is there any scientific value to having a man fly to the moon and back?" NASA suggests that Webb respond as follows: "I am convinced there is. I recently heard one well qualified scientist say that a geologist spending thirty minutes on the moon would bring back more information than we would gain by placing 100 instrument packages on the moon."[30]

In his testimony to a congressional committee, Webb described the American space effort as one "designed to accomplish those things which are most in our interest scientifically," while NASA's General Counsel John A. Johnson called the program "justified by nothing more tangible than the total national interest in maintaining leadership in science and technology."[31] Yet justifying a manned lunar space mission through scientific advancement presented problems. Numerous respected scientists publicly challenged the scientific benefit of a manned lunar landing. Early in 1961, a senator asked Hugh Dryden whether a permanent manned space vehicle would not gain more information scientifically than a moon shot. Dryden responded that NASA did not justify manned space flight scientifically, insisting that it was "mainly a technological undertaking."[32] By 1963, scientific criticisms had become louder and more frequent. When queried about the scientific benefits, Dr. Homer E. Newell, director of the Office of Space Sciences, explained that the United States gained a "collection of things" from the program; "not any one of them justified the program."[33] Finally, in response to a question about moon exploration, Webb conceded that a U.S. lunar landing would be short and could not support "a significant scientific exploration of the moon."[34] Fortunately, NASA had a variety of other arguments to rely upon when the scientific arguments proved unpersuasive.

Anticipating arguments from social liberals, the administration proposed that a manned moon mission would benefit the country educationally. Science and education would be "enriched by knowledge of our universe and environment," President Kennedy proclaimed, "by new techniques of learning and mapping and observation, by new tools and computers for industry, medicine, the home as well as the school."[35] Vice President Johnson spoke out too, proposing that the space age was causing "revolutions" in American education.[36] James Webb perhaps made the best pitch: "We hope to support and help the schools and colleges in their efforts to build strong basic educational programs that are essential to our national progress."[37]

Touting the educational benefits early in the program helped to lessen criticisms surfacing in 1962 and 1963 that money devoted to a manned space shot would be better spent on education, housing, and other social programs. In a press conference in 1963, Kennedy added a caveat to criticisms that space money would be better spent on "housing or education." Congress, Kennedy asserted, would "cut the space program and you would not get additional funds for education." Later, when the Soviets made another space spectacular, he continued, people would ask, "why didn't we do more?"[38]

The administration had to perform a tightrope act in attempting to justify Project Apollo militarily. A military justification appealed to various conservative members of Congress. Yet Eisenhower created NASA as a civilian organization aimed at the peaceful exploration of space. He did so, in part, to end the fighting among the service branches for control over space exploration. The Kennedy administration attempted to separate NASA rhetorically from the Soviet Union's space program by highlighting NASA's peaceful intent. The agency's public relations department highlighted its peaceful intent by creating the National Conference on the Peaceful Uses of Space.

Vice President Johnson spoke often about America's peaceful intentions. At a space center dedication, he declared: "Our national purpose in space is peace." Space successes, he continued, may bring closer "a world of universal peace, freedom, and justice."[39] General Bernard Schriever concurred, characterizing America's intentions in space as peaceful. Space power, he added, is "peace power."[40] While touting peaceful intentions, the administration never relinquished popular, persuasive appeals to national defense. Representatives argued that the program's purpose was peaceful *and* would strengthen the military role in space. George Low remarked that NASA was "interested in getting the most out of the space program both for national defense and for peaceful applications."[41] General Schriever echoed Low: "Only by being strong can we preserve the peace."[42]

Throughout 1961 and 1962, administration officials spoke of the defense needs met by the space program. President Kennedy warned of the threat to the national security posed by Soviet space accomplishments. America cannot permit the Soviet Union to "dominate space," he cautioned, "with all that it might mean to our peace and security."[43] This country, the president warned, simply cannot afford to be second in outer space with its "many military implications" still "unknown."[44] In 1963, Vice President Johnson reassuringly proposed that America would not accept an inferior place to the Soviets in space. "Americans," he said at a luncheon in honor of Gordon Cooper, "do not intend to live in a world which goes to bed by the light of a communist moon."[45] Department of Defense (DoD) representatives, hesitant to become too involved because of its own programs, did at times back up NASA's contentions. General Schriever told a House subcommittee that Project Apollo supports American "national security needs." Militarily, explained Schriever, "the optimum system in space will utilize man."[46]

Confronted with increased criticism of a manned lunar shot in 1963, NASA officials leaned heavily on the military justification. Testifying before the Senate Appropriations Committee, Webb reinterpreted America's initial purpose for entering space exploration. America, instructed Webb, embarked on the program in "the face of a new Communist threat to our national security."[47] D. Brainerd Holmes, director of the Office of Manned Space Flight, provided a similar response when queried about the scientific benefits of a manned lunar shot. Sidestepping the scientific justification, he suggested that NASA had designed the program "to advance the welfare and security of the Nation."[48] Webb, in particular, continually mentioned the potential national defense benefits in his testimony before congressional committees, commenting on the indispensability of large booster rockets to military power and proposing that NASA's Saturn complex could be "taken over" immediately by the military if the United States needed to conduct military space missions.[49] Not all DoD officials provided unequivocal support for the military applications of manned space flight. Testifying before the House Committee on Science and Astronautics in 1963, John Rubel, the assistant secretary of defense, said he did not think the Defense Department had made "any judgment at all with respect to the necessity or nonnecessity of man" in space.[50]

NASA grew quickly under the Kennedy administration, especially in 1962 and 1963 with the acceleration of manned space flight. Administration officials attempted to justify Project Apollo by pointing to the knowledge gained from managing such a large project. Joseph F. Shea, director (systems), Office of Manned Space Flight, called Apollo "a yardstick" for measuring America's ability to "manage a great engineering and technological undertaking in the national interest."[51] D. Brainherd Holmes echoed Shea's claim: "The national experience of managing a research and development program of this magnitude is an added benefit." In the future, he projected, we will be able "to undertake even larger programs."[52] Finally, in presenting awards to NASA administrators, Webb called the men who had the ability "to organize" the worldwide Mercury program "those who really count."[53] The large programs NASA organized cost tremendous sums, and administration officials attempted to justify the agency's expenditures in numerous ways.

Administration officials attempted to appeal to business leaders and the average American by arguing that space exploration would benefit industry and improve Americans' standard of living. NASA's public relations department anticipated questions about exploration's high cost. It suggested that Webb respond to inquiries about whether Americans would support NASA's space programs "money-wise" by pointing to the increased "standard of living" derived from the spending.[54] Acknowledging space exploration's high price tag, President Kennedy claimed it would have a "far-ranging effect within industry."[55] Hugh Dryden echoed Kennedy, stating that improved "industrial processes" would raise Americans' "standard of living."[56] The vice president went a step further, characterizing the space age as "a second industrial revolution."[57]

The vice president and various NASA officials asserted that space spending would stimulate the economy. Johnson proposed that the program would sharply "stimulate" the country's "productive output," leading to increases in "our gross national product, our income, and the Federal Government's tax intake."[58] Once James Webb went so far as to imply that the health of some American businesses had become inextricably tied to NASA's funding. During the heat of a Senate appropriations battle in 1963, Webb justified the program with a hint of economic blackmail. He responded to a senator's question about what proposed House cuts would do to NASA: "I really hesitate even to start discussing possibilities that affect commercial companies that might be reflected in the quotations on the New York Stock Exchange."[59]

Administration officials also tried to deflect criticism of NASA's enormous budget. It is "imperative" that America not set its space objectives by "the narrow horizons of a dollar debate," argued the vice president. The current "preoccupation" with costs stems from the inability of the political and technical communities to explain the program.[60] Officials also compared space spending with other levels of spending to demonstrate that the cost of a manned lunar landing was not exorbitant. When asked about the price tag of putting a man on the moon, Hugh Dryden pointed out that the cost corresponds to what the nation spends "in the Defense Department in a year and a half."[61] Vice President Johnson went a step further, pointing to the high expenditures for unnecessary items. America spends more on "cigarettes," "alcohol," "recreation," or "horse racing" than it does in space.[62]

NASA attempted to demonstrate that the space program benefited business while simultaneously trying to dispel negative impressions of governmental bureaucracies and space exploration. NASA's public information office, in a memorandum to Webb, anticipated questions about how much of the agency's budget would be spent "inhouse" and how much would "go to industry." The public information office informed Webb that NASA spent 85 percent to 90 percent of its budget for "services outside the government."[63] During space hearings in 1962–1963, Webb referred often to the high percentage of NASA's funds spent in contracts with "corporations," "university teams," and "nongovernmental entities." One finds identical testimony offered by other NASA officials. Interestingly, the figure rarely came out in response to a direct question. Officials usually provided the material as part of their presentations. Besides attempting to dispel NASA's image as an expensive bureaucracy, some administration officials occasionally reminded members of Congress that funds would be spent on earth, not in space. Dryden explained that the dollars NASA requested would not be "fired to the Moon" but would be spent in "factories and universities" here on Earth.[64] NASA spends its money "on the ground," Webb observed, but equipment will do the work "out in space."[65]

One should view space exploration not solely as an immediate economic catalyst, officials explained, but also as a long-term investment. James Webb attempted to portray space spending as an investment in the country's infrastructure. NASA's funds, observed Webb, will go toward "construction," "maintenance," and "repair for this country."[66] Vice President Johnson also attempted to portray space spending as a long-term investment. In a speech at the Second Conference on the Peaceful Uses of Space, he termed the space program a "profitable investment—not a waste of the Nation's resources."[67] Conservative estimates, Johnson exclaimed in another address, project that "outlays will yield $2 return for every $1 invested."[68] Later challenged by critics, NASA was forced to abandon the claim.

A stimulated economy, the administration argued, meant more jobs. In late 1962, President Kennedy declared that the space effort, still in its infancy, had already created "tens of thousands of new jobs."[69] Calling Project Apollo a "thriving enterprise," Wernher von Braun, director of the Marshall Space Flight Center, observed that it employed "hundreds of thousands" of Americans. If Apollo's budget were cut, von Braun warned, NASA would have to lay off numerous workers, causing reverberations "throughout the Nation."[70] Similarly, James Webb argued that expansion of the space industry would create "new jobs," picking up the "slack caused by automation."[71] Ironically, NASA justified its programs, in large part, in terms of the technological revolutions it would produce in science and industry. Yet the improved technology would lead to greater automation and greater displacement. Finally, the vice president tried to characterize the space effort as a force for social justice. The space age necessitated hiring the country's best talent, "regardless of race or religion." Thus, he asserted, the space effort would help sweep away "senseless patterns of discrimination in employment."[72]

Appealing to the American consumer, administration officials pointed to the various by-products created from space research, from ice cream to steel. Americans, the president asserted in 1961, will feel the impact of space science in their "daily lives."[73] The space program had tremendous potential for creating new consumer products. In 1961, Webb told an audience that the program had created more than "3,200 space-related products."[74] Vice President Johnson, moreover, proposed that every American should ask what the space program would mean to him or her in "tangible terms."[75] The by-products had practical applications, submitted Johnson. Americans would soon see space "by-products" in their "automobiles," their "office equipment," their "air conditioning and heating," and even in their "pots and pans."[76] NASA's deputy administrator, Hugh Dryden, also highlighted the benefits to the average American, stating that the space effort would produce a tremendous variety of consumer goods. Unfortunately, Dryden explained, Americans will not always recognize the space program as the "origin" of many future products.[77]

Critics of Project Apollo did not have the same reasons for opposing it. Although some criticized a moon shot as wasteful, a great many others objected only to *manned* flights. Unmanned flights, they contended, could accomplish the tasks of early exploration more easily and more cheaply than manned flights. The administration, however, viewed manned flight as a necessity in light of the propaganda battle with the Soviets. The Kennedy administration was concerned with national prestige. Even though an unmanned flight might produce an equal or greater quantity of useful information, manned flight would command the attention and capture the imagination of an impressionable world. Clearly, the administration believed, the world would view the nation best able to execute manned missions as the space leader.

Administration officials attempted to justify man's presence in space by highlighting unique human qualities: sight, judgment, analysis, reason, flexibility, reliability, and manual dexterity. Although one finds some officials expounding man's virtues directly, one finds most of the justification in prepared written material submitted to space committees and subcommittee members in response to questions about specific programs. Only when man enters the space environment will one realize the "most significant achievements," begins one prepared statement.[78] Another proclaimed that space exploration in its "truest sense" depends on man's direct participation.[79] NASA relied most heavily on two arguments: man's flexibility would allow him to respond to "unforeseen situations," and man can use his brain to analyze problems and make instant decisions.

Obviously, NASA attempted to stress man's unique possession, his ability to reason. One NASA document characterized man's brain as the "ultimate reservoir and integrator of all human knowledge," while Alan Shepard called man "a computer" that, per pound, works more effectively than "any mechanical computer."[80] Amid the criticisms of the space effort in 1963, President Kennedy made his only comment on why *man* should go into space at the presentation of NASA's Distinguished Service Medal to Gordon Cooper. The president called man "the most extraordinary computer of all." Echoing NASA's earlier statements, Kennedy added that man's "judgment," "nerve," and learning capability make him "unique." These qualities necessitate continuing "manned flight."[81]

Not all NASA representatives believed manned flight was the only means of meeting the country's space objectives. Responding to the suggestion that his testimony implied that NASA could gather data more easily with a man in space, Dr. Abe Silverstein, director of the Lewis Research Center, agreed that part could be obtained more easily with a man in space. He admitted that part could be "obtained without a man."[82] With the increased criticism in 1963, some members of Congress questioned NASA officials about the usefulness of unmanned flights. Dr. Homer Newell, director of the Office of Space Sciences, answered a query about the amount of scientific information one could obtain from an unmanned lunar shot; he revealed that NASA could obtain "a very large fraction" of its scientific information with an "unmanned" shot.[83] Responding to a similar line of questioning, Webb admitted that "exploration by unmanned means could probably also solve most of the major problems." But, he asserted, "the effort and funds required would not be insignificant."[84] In short, NASA officials had to admit that unmanned lunar shots would have obtained the needed information—and probably at a fraction of the cost of a manned moon mission.

In attempting to sell a manned moon shot to Congress and the nation, the Kennedy administration developed a large, active public relations program. The space agency's PAO worked hard to project a positive image. It courted media coverage in numerous ways. It used the media's coverage of the president to gain publicity; it staged or sponsored events; it controlled all information coming out of the agency, especially the remarks of the astronauts; it entered into contracts with media capable of spreading its message; and, although there is no concrete proof, the agency may have used the timing of its activities to enhance its chances of gaining funding for its programs. The public relations strategies and tactics were aimed not only at gaining headlines but also at spreading its justification for space exploration.

The agency, and the administration, presented numerous arguments favoring a manned lunar landing, placing greater emphasis on different arguments over time. The administration argued that the nation must proceed with a manned lunar landing to maintain its image as the world leader, to advance science and technology, to improve education, and to gain valuable experience managing large projects. Space exploration, the officials argued, had both peaceful and military purposes. The administration presented a consistent picture of the positive benefits a manned moon mission would have for the economy. Not only would it stimulate the economy, improve the standard of living, and create thousands of new jobs and consumer products, but it would also serve as an investment in America's future. Finally, man must go into space, the administration declared, because of his unique human qualities. In justifying Project Apollo, the administration attempted to appeal to everyone, from the dove to the hawk, from the educator to the businessperson, from the homemaker to the homeless. Even though the administration admitted that it could probably gain many of the same results from unmanned flights, they argued that man's unique characteristics were necessary to get the most out of the exploration of space.

While all these arguments may have been persuasive, they constitute only part of the Kennedy administration's case. The administration also depicted the manned lunar landing in narrative form as a great frontier adventure, complete with heroes and villains. Although critics would question the political, scientific, military, and economic justifications for sending a man to the moon, the frontier narrative went unchallenged. Both the media and Congress found the story irresistible. In short, the frontier narrative stood as the most powerful justification for a manned moon mission.

2

The Kennedy Administration
and the New Frontier

The importance of Lewis and Clark's trailblazing on the western frontier, writes Henry Nash Smith, "lay on the level of imagination: it was drama, it was the enactment of a myth that embodied the future."[1] One can say the same of America's trek into the New Frontier of space. Americans demonstrated their fascination with a manned moon mission, encouraging Congress to spend enormous sums of money on the program. In the early 1960s, as historian Walter McDougall has written, Project Apollo was "the greatest open-ended commitment by Congress in history."[2] To understand this commitment, and the public fascination behind the space program, one must go beyond the political, scientific, military, and economic arguments of the Kennedy administration's public relations campaign to consider how the administration dramatized Project Apollo as a great frontier adventure.

Human beings, storytelling animals, perceive and give meaning to actions through narration. "Not only do we understand our own actions in terms of narrative structure," as Janice Hocker Rushing writes, "but we find purpose and guidance for our lives in accord with the stories told by the society in which we live."[3] Cultures return to their stock of stories, or myths, as they interpret the meaning of actions and events. Among the most important of these stories, particularly in the American experience, is the myth of the old frontier. "The essence of all that is genuinely exceptional in American history," notes Richard Slotkin, "is embodied in those myths that are peculiar to our culture, of which the oldest and most central is the myth of the frontier."[4] Throughout its history, as Rushing writes, America has relied heavily on the frontier for its "mythic identity." From its initial settlement to its conquest of the western wilderness, America has drawn upon the frontier myth as a source of identity and as a motive for action.

The frontier myth is a distinctly American adaptation of the traditional hero story. Although the frontier myth may differ from the traditional hero story in certain details, the two have in common an essential function: to convey how heroes develop an awareness of their strengths and weaknesses in a manner equipping them for the difficult tasks they will face in their lives. The task of the ego-conscious hero is "to achieve independence from unconsciousness and assert control over it; this battle is typically expressed as a triumphant struggle with the forces of evil."[5] Although superior to the common person in many ways, archetypal heroes are susceptible to the sin of pride (hubris), or of mistakenly equating themselves with the gods.

The American frontiersman shares many characteristics with the archetypal hero of earlier myths. Like the traditional hero, the frontiersman had evil forces to contend with, both a hostile, unknown environment and the sinister inhabitants lurking within it. The American frontier myth features a rugged, independent pioneer who attempted to conquer the land and its inhabitants, thereby expanding the country's domain and improving its way of life. Using sheer willpower to control his ego and preparing himself for the battle, the frontiersman "demonstrates no emotion, and through practice with weapons, he readies himself for the fight." But the frontier hero, conspicuous by his individualism, frequently becomes alienated from the community he battles to save. Rushing proposes that this tension between individualism and community, "more than anything else, defines the Old West."[6]

President John F. Kennedy evoked the spirit of America's frontier mythology as he integrated his space policies with the general theme of his administration: the New Frontier. In an address on 25 May 1961, which Ralph G. Martin deems the "real birth" of America's space program, Kennedy announced America's commitment to land a man on the moon before the end of the decade, calling it the "exciting adventure in space."[7] During telephone remarks to NASA's first Conference on the Peaceful Uses of Space, Kennedy expressed America's determination to continue to be "a pioneer in the new frontier of space."[8] The romance and intrigue of the moon, coupled with the mythology of the frontier, gave Kennedy a way of depicting a march to the moon that was both exciting and concrete.

President Kennedy's New Frontier, writes David Zarefsky, "became a meaningful symbol when it received widespread use and when the related images of discovery, exploration, charting a course, and pursuing the unknown were given expression."[9] And nothing seemed capable of giving these images more powerful expression than Kennedy's rhetoric on the space program.

According to Rushing, the attempt to relocate the American frontier in outer space changed the basic structure of the frontier myth.[10] She identifies two major differences between the old frontier and the new one. First, unlike pioneers of the Old West, the astronauts could not conquer their new environment, because outer space was limitless or infinite. Second, technology, or the means of traveling into space, by becoming of paramount importance, reduced the astronauts to mere passengers. The astronauts, encapsulated in their crafts, could not act—could not exert control—like pioneers of old. According to Rushing, these differences exemplify an evolutionary change from the old frontier to the new, transcending the distinctions between community and individuality, harmony and conflict.

Rushing's analysis of the relationship between the frontier myth and the space program has provided useful insights into the structure of the frontier narratives and the reasons for their persuasiveness in the American culture. But she says little about the actual impact of the frontier motif on support for the early American space program, for she analyzes, not the rhetoric and events of the time, but a dramatized account of the space program written more than a decade and a half later: Tom Wolfe's *The Right Stuff*.[11] As a consequence, Rushing's analysis has little direct relevance to political debate over the space program in the early 1960s; if it has any political relevance at all, it suggests how Wolfe's rendition of the space program might have helped shape political events in the 1980s (such as support for the space shuttle program or John Glenn's candidacy for the presidency). It remains to assess the character and implications of frontier narratives in the actual political debates over NASA's manned space program from 1961 to 1963.

Space: The Final Frontier?

President Kennedy himself led the administration's effort to dramatize the space program as a great frontier adventure. Addressing the employees of the McDonnell Aircraft Corporation in St. Louis, Kennedy characterized the shot as "the most important and significant adventure" in the history of the world.[12] In his speech at Rice

University in late 1962, Kennedy called Project Apollo "the most hazardous and dangerous and greatest adventure on which man has ever embarked."[13] He made a similar comment while presenting NASA's Distinguished Service Medal to Gordon Cooper in mid-1963, when he described the space program as the "great adventure of the sixties."[14] NASA officials echoed Kennedy's depiction. George Low, director of Spacecraft and Flight Missions, Office of Manned Space Flight, spoke of "sharing the adventure" of space exploration while John Johnson, NASA's General Counsel, referred to it as "a great new human adventure."[15] Americans thus learned to view a manned lunar shot, not merely as a scientific enterprise, but as a great human adventure.

The trip to the moon was to be more than just an exciting adventure. NASA officials also characterized Project Apollo as fulfillment of man's destiny, much as their forefathers had spoken of America's destiny to explore and expand the western frontier. In a speech to the New York Patent Law Association entitled "The New Frontier of Space," John Johnson characterized man as an "explorer by nature" who "demands the unknown be reduced to the known."[16] Man, in short, was compelled by a thirst for knowledge to investigate the mysteries of space. During an awards presentation, NASA administrator James Webb proposed that man faces a new space age—"the age man equally accepts his destiny to advance out into space."[17] Robert Seamans, NASA's associate administrator, offered similar testimony. "Man," he asserted, "is destined to play a vital and direct role in the exploration of the Moon and the planets."[18] Just as Americans had portrayed earlier frontier forays as part of America's destiny to extend freedom in a world full of tyranny, the unlimited bounds of space now constituted what Vice President Lyndon B. Johnson dubbed the "New World of freedom."[19]

The Kennedy administration frequently relied upon direct analogies to exploit the powerful myth of the Old West. "What was once the furthest outpost on the old frontier of the West," the president proclaimed, "will be the furthest outpost on the new frontier of science and space."[20] Kennedy made even stronger comparisons during 1963, proposing that space exploration brought back a rugged, pioneer spirit. Presenting NASA's Distinguished Service Medal to Gordon Cooper, the president took the opportunity to mention that in America's "rather settled society," the astronauts had "demonstrated that there are great frontiers still to be crossed."[21] In a speech intended for delivery in Austin, he cautioned that space was "still a daring and dangerous frontier." Texans had battled foes on the frontier before, Kennedy recalled, and they would help him "see this battle through." Kennedy called it a special time in America's history, "a time for pathfinders and pioneers."[22] Vice President Johnson also compared space exploration to a trek into a frontier. At a space center dedication, he called space mankind's "last and greatest frontier."[23] Like Kennedy, Johnson drew a direct analogy between space and the Old West; "We go into space as pioneers came into the West, for one purpose only:" to discover a better life and to secure our freedom.[24] Just as the Old West had allowed man to live independently and ensured his freedom, so would space.

The astronauts, of course, played a special role in the unfolding drama. As the principal characters, their words carried special significance. Many of their comments echoed those of the administration. Scott Carpenter described space as "a fabulous frontier," while Alan Shepard explained that he joined the program out of an "urge to pioneer."[25] John H. Glenn, Jr., and Virgil I. "Gus" Grissom, like administration officials, compared exploration of the western frontier and exploration of space. Glenn described space flight as the "great exploration of all time."[26] The astronauts, Glenn proposed, felt privileged "to serve as the pioneers" of the program.[27] They tried "to blaze a trail" for those who would follow them.[28] Like Glenn, Grissom wrote of "a spirit of pioneering and adventure." Grissom went even further, however, stating that had he lived 150 years earlier, he might have wanted to "help open up the West."[29]

Elements of the Frontier Adventure

To ring true, a frontier story must possess specific constituent elements: (1) an identifiable, conquerable geographic location that is (2) unknown and hostile and includes (3) a malevolent antagonist who is thwarted by (4) a heroic adventurer. As Rushing has suggested, space exploration inherently differs from the westward journeys of the real pioneers in at least two important respects: the nature of the "scene" and the role of the "hero."

Central to a frontier adventure is an appropriate "scene"—an identifiable, geographic location to conquer and dominate. Outer space, intangible, infinite, and therefore unconquerable, as Rushing has noted, did not fit the traditional notion of a frontier. Rushing writes that actually space is "infinite." Unlike the "Old Frontier," one cannot eventually conquer or fill up space, "for it has no boundaries. Indeed, the term 'space as scene' is an oxymoron, for 'scene' is inherently a material term."[30]

In the Kennedy administration's space narrative, however, this difference never proved a significant problem, for the adventure *did* have a concrete, objective goal: landing a man on the moon. Wernher von Braun, director of the Marshall Space Flight Center, explained the political and dramatistic importance of having Kennedy's "crystal clear" goal of landing a man on the moon before the end of the decade.

Everyone knows what the moon is; everyone knows what this decade is; and everyone can understand an astronaut who returned safely to tell the story. An objective so clearly and simply defined enables us to translate the vague notion of conquering outer space into a hard-hitting industrial program that can be orderly planned, scheduled, and priced out. It establishes a sorely needed, firm, nonvacillating goal which alone can serve as a basis for a long-range plan.[31]

As von Braun points out, instead of the vague notion of conquering a limitless, infinite outer space, a moon landing provides a concrete objective. Although one might doubt the economic or scientific value of landing a man on the moon, there is little doubt that it had great rhetorical value in romanticizing and concretizing the "frontier" of space.

With a tangible and, presumably, worthy objective established, administration officials could emphasize the hostile, unknown character of the New Frontier to justify both their failure to "keep up with the Russians" and the need for more funding. Trying to appease anxious members of Congress after Yuri Gagarin's orbital flight, James Webb rationalized the comparatively unspectacular progress of the American program by reminding his audience that in the "unknown area" of space, one finds "hazards." For man, he continues, space is a "hostile environment."[32] Webb returned to emphasizing the "hostility" of space during an appropriations meeting in 1963. Seeking additional funding for NASA's advanced research facilities at Ames, Webb asserted that the "hostile environment" of space necessitated such expenditures. The following day, he submitted additional material justifying increased expenditures that described space as a "foreign and hostile environment."[33] The theme of space's "hostility" assumed special emotional force when the astronauts themselves—those who would risk their lives—joined NASA administrators in stressing the hostility of space. In *We Seven*, the astronauts' chronicle of the space program, John Glenn and Alan Shepard also described space as hostile. Glenn wrote of the "hostile elements of space," while Shepard described it as "one hostile environment."[34]

How could one call space hostile? Hazardous, perhaps, but hostile? Hostile seems to imply antagonistic actions or attitudes—an enemy with malicious intent. The wilderness of the traditional frontier had wild animals one might view as hostile. More important, it had the Indians, whom frontier storytellers often portrayed as brutally savage. Space, on the other hand, was devoid of life; it presented only passive and inanimate hazards, like radiation or lack of oxygen. In both "frontiers," man had to battle the forces of nature. A crucial element on the frontier scene seemed to be missing from the "frontier" of space: a tangible, human villain creating obstacles for the hero to overcome.

Thus the Soviets came to play a crucial role in the New Frontier narrative. Although the administration did not always refer to the Soviets by name, they were clearly the malevolent antagonist, much like the Indians of the old frontier. The Soviets, moreover, did not merely seek to defend their native lands, like the Indians of the old frontier; they sought to conquer the world. Therefore, President Kennedy warned that the United States could not permit any nation "to dominate space" whose "intentions" toward it "may be hostile."[35] Vice President Johnson metaphorically described freedom as a "sturdy plant" but cautioned that freedom cannot "grow and flower" on earth when the universe enveloping it is "poisoned and contaminated by tyranny."[36] Much like cavalry officers explaining possible Indian attacks, American military officials, such as General Bernard Schriever and General Curtis Lemay, also referred to the Soviets as a "hostile competitor," warning of "the growing Soviet space threat" to American security.[37] NASA administrator James Webb used the harshest language, however, calling the Soviets "a powerful despotism, bent on burying us along with the basic tenets upon which our society rests and from which it draws its strength."[38] How they hoped to use space to accomplish this feat Webb never made clear.

Even more important to the frontier narrative than an appropriate scene and a hostile enemy was a suitable hero: the brave frontiersman. The frontiersman in space had to embody what Americans liked to believe were traditional American values, combining traits of both the Puritans and the pioneers. From the Puritans, one would expect qualities like humility, discipline, and religious devotion. Puritans preached self-control, control of the human appetites and emotions that might lead one astray. Like a boxer preparing for a fight, a true frontiersman would control his sexual urges, forgoing intimacy during his preparation and journey. From the pioneers who settled the western frontier, the new frontiersman would learn courage, patriotism, and fierce self-reliance. In short, the hero of the New Frontier would have to struggle to control himself and his consciousness. To do so, he had to purge himself of emotion; reason must prevail.[39]

The pioneer of the space frontier had to be part Davy Crockett, part Buck Rogers. He had to possess not only the traditional pioneer qualities but also the new technical expertise needed to operate in space. From the start, the seven astronauts seemed to have many of the needed attributes. As military men, they exemplified the tension between individual and community. They worked, lived, and even dressed in ways that separated them from the larger community. Moreover, the astronauts came from an isolated and select group of military men; they came from the ranks of test pilots who lived in special military bases set on barren stretches of land. The spartan living conditions, the insufferable climate, and the day-to-day flirtation with death eliminated all but the most rugged individualists. And finally NASA isolated this elite group even further, selecting only the top seven test pilots and cloistering them for physical and psychological examinations.

The astronauts themselves were quick to reinforce their images as unique individuals in the mold of the traditional frontiersman. As test pilots, they reminded the public, all of them had demonstrated both the courage and the skill to face the unknowns awaiting them on the New Frontier. Experienced test pilots, Deke Slayton observed, "run into things no one has yet written a book about."[40] Scott Carpenter observed that the test pilots had benefited from harrowing experiences, which built up their tolerance of fear and their "ability to face the unknown."[41] John Glenn provided almost identical testimony, insisting that he felt qualified for the program because he had a lot of experience facing "dangerous unknowns."[42]

A crucial difference, however, imperiled the analogy between space exploration and the old frontier. Pioneers of the past conquered their respective frontiers by actively exerting control over their "unknown," hostile environments. Frontiersmen of the Old West used rifles, axes, and knives, along with their reasoning capabilities, to thwart their foes. Sea captains navigated their ships through uncharted waters by their own intuition and skill. Even the early test pilots controlled their own adventures into "the wild blue yonder." Through their actions, all these earlier "heroes" controlled their own destinies. For the frontier story in space to ring true, the astronauts could not be mere passengers; they too had to appear to exert control during their flights.

The fact was that the early astronauts were merely passengers. As Rushing notes, the astronauts "were not using technology to earn their badges. It was using them. They were literally en-capsule-ated by it . . . and not even able to watch the process of their own manipulation."[43] Man served merely as backup. Indeed, even animals could "pilot" the flights, as NASA demonstrated on two occasions. Four months before its first manned suborbital flight, NASA shot a chimpanzee named Ham 157 miles into the sky—40 miles higher than the first two manned flights. Another chimp named Enos completed a dual orbit of the earth three months before an American performed the feat. The astronauts could not "control" the orbital path or the speed at which the capsule traveled, only the angle at which it floated—sideways, backward, forward. This hardly seemed an appropriate role for a heroic adventurer.

Even the astronauts themselves initially expressed doubts that Mercury flights required anything more than human guinea pigs. Walter Schirra expressed reluctance at throwing away years of flying to participate in what "sounded like a stunt."[44] Gus Grissom echoed Schirra's skepticism, disclosing that the program sounded too much like a "stunt" in search of a "passenger."[45] Deke Slayton evinced a similar reaction, recalling that at first he thought the program really needed not "trained test pilots" but a "human body" to tie to a "missile" and fling into space.[46] Even after they had joined the program, the astronauts occasionally undermined the effort to portray them as crucial to the success of the program. At a press conference in 1959, Alan Shepard admitted that the Mercury flights were "preprogrammed and autopilot, with the pilot's manual only as backup."[47] Walter Schirra pointed out that the astronaut could not change the "orbital path" but could merely maneuver in it.[48]

More often, however, the astronauts promoted the idea that only a select few had the qualifications to "fly" in the space program. Walter Schirra argued that the project required "really good test pilots" to "handle the job."[49] Gus Grissom insisted that the flights required a trained pilot and that his initial impressions were wrong.[50] Deke Slayton announced that after NASA's initial briefing, he realized only a "test pilot" could "hack this flight."[51] Maybe NASA's briefings did indeed persuade the astronauts of their important role in the flight, or perhaps they simply learned the importance of public relations. Whatever the reason, the astronauts ultimately led the way in portraying themselves as "in control."

NASA officials reinforced the astronauts' claim of control over their missions in subtle yet crucial manipulations of language. Conveniently, NASA could call the astronauts pilots because of their backgrounds. The label "pilot" conveys the notion of control, and NASA never lost an opportunity to use the term "pilot" when referring to the astronauts. In a 1959 press conference, reporters asked George Low why the astronauts received so much training when their flights would be "limited" and "thoroughly automated." Low responded, "Even though the flight is completely automated, we feel that the Mercury capsule is still essentially a flying machine." If something goes wrong, Low added, the astronaut "would have the possibility of navigating, of controlling the attitude, of taking all the emergency procedures himself." The intensive training the astronauts receive, Low asserted, would help them "do a much better job of flying."[52]

NASA sensitivity to the pilot-passenger distinction may also have led to a subtle change in the label attached to the astronauts' space vehicle. In 1959, NASA called its Mercury vehicles "capsules." Sometime between September 1961 and January 1962, NASA erased the word "capsule" from its vocabulary and replaced it with the word "spacecraft."[53] The word "spacecraft" first surfaced during a 1960 budget rehearsal, when a NASA official used it with reference to large "lunar landing vehicles."[54] Prior to John Glenn's flight, NASA began calling his vehicle a "Mercury spacecraft." Rhetorically, "capsule" and "craft" seem a world apart. "Capsule" implies something sealed, encapsulated, uncontrollable. A craft, on the other hand, plainly recalls navigable boats that one can pilot.

Administration officials sometimes unwittingly betrayed the characterization of the astronauts as pilots. Late in the astronauts' initial press conference, for example, George Low acknowledged that in orbital flights, NASA would retain control: "We will have the possibility of course to bring him down after one or two orbits." Interestingly, however, he quickly shifted the attention back to the "control" of the astronaut, adding: "or he [the astronaut] will have the possibility to come down."[55]

Even Low's "slip" thus ultimately provides additional evidence of NASA's concern with maintaining an appropriately heroic image for the astronauts. Combined with the administration's portrayals of space as a hostile environment, its emphasis on the threat posed by the Soviets, and the individualism of the test pilots, the administration's suggestions that the astronauts controlled their own destiny completed a structurally coherent narrative of space as the New Frontier. Unfortunately, characteristics of some of the flights and also of the men aboard them on occasion made it difficult to construct frontier narratives that seemed to ring true. In short, the *situation* sometimes created problems for the New Frontier narrative. An examination of the early flights of Project Mercury illustrates how the actual events and personalities of the space program sometimes made it very difficult to "sell" space as a New Frontier.

Situational Constraints and the Frontier Myth

The first two Mercury missions generally failed to evoke the excitement of later flights, and a significant part of the reason may lie in failures related to the frontier mythology. From the very first flight, the administration attempted to invite an analogy between space and the old frontier. But for a number of reasons, some errors of rhetorical strategy and others involving the character of the flights and the astronauts themselves, the administration's attempts to generate enthusiasm for the first two flights did not succeed.

In characterizing Alan Shepard's suborbital flight of 5 May 1961, the administration heavily emphasized the notion that the astronauts, like frontier heroes, controlled their destiny through their skill and daring. A press release issued one week before the flight emphasized how the astronaut would demonstrate "manual control of the spacecraft attitude before, during, and after retrofire."[56] Postflight press releases likewise stressed the necessity of having a man aboard the flight and told how Shepard had controlled the system manually, one axis at a time. "This was done because a pilot had never controlled a craft in space before,"[57] NASA explained. Shepard's own statements also emphasized how he "had full control of the craft."[58] The astronaut even exploited the issue of "control" to deprecate the "enemy." Soviet Major Yuri Gagarin had a "fine long ride," Shepard observed in a *Life* article, but "he was a passenger all the way."[59]

By so explicitly stressing the matter of "control," the administration actively focused attention on the greatest difference between the traditional frontier myth and the realities of the space program: the responsibility of the "hero" for his own destiny. In addition, the flight itself, especially in comparison to the Soviets' orbital flight, seemed simply too unspectacular to "sell" as a great adventure into the New Frontier. The flight consisted of Shepard and his capsule being shot like a bullet 116 miles up into the sky and then dropping back down into the ocean. The entire flight lasted only fifteen minutes. Comparing such a short, up-and-down ride with the trailblazing of Davy Crockett stretched the bounds of imagination.

The decision not to engage in a major campaign of preflight publicity also hindered the administration's ability to evoke the frontier mythology. NASA did not build up Shepard as a "hero" about to embark on a frontier adventure; indeed, the agency even refused to announce prior to the flight that he would serve as the first astronaut. His name was released only after his launch had been canceled at the last minute. When launched three days later, he still ventured into space a relative unknown. NASA's silence about Shepard is understandable. After broadcasting to the world numerous fiery failures by unmanned rockets, NASA may not have wanted the public to become too attached to Shepard. With little knowledge of the man, Americans probably found it difficult to fit him into the heroic mold. Moreover, Shepard himself seemed to shun the "hero" image. " 'We were asked to volunteer,' " Shepard once remarked, " 'not to become heroes.' "[60]

Finally, Shepard's flight had no clear objective—no ultimate "scene" to conquer. Kennedy made his speech committing America to a moon shot two weeks after the flight. Without an identifiable, concrete goal like the moon, the parallel between the western wilderness and outer space seems less believable. Before Kennedy's speech, NASA's flights seemed to be headed nowhere in particular. They simply touched the boundaries of infinite outer space. After the speech, the flights became at least small steps toward a concrete goal: landing a man on the moon.

The second Mercury flight, "piloted" by Gus Grissom, thus had at least one advantage over the first. It was to be the first step in the great adventure of landing a man on the moon. Yet numerous other factors once again limited its potential for success as a great frontier adventure. Like Shepard's flight, the second mission seemed simply too unspectacular to stir memories of the great pioneers. Grissom's flight was just another up-and-down affair. As with Shepard's mission, the agency did not disclose the identity of the astronaut until a day before the scheduled launch, thus precluding any buildup of anticipation and excitement among the public. And like Shepard, Grissom rejected attempts to portray him as a hero. " 'I'm not the hero type,' " he confessed prior to his flight.[61]

Grissom's flight on 21 July 1961 certainly seemed to confirm his lack of heroic qualities. His launch went off without a hitch. But when he landed in the ocean, the escape hatch on his craft mysteriously opened, and the expensive craft sank to the bottom of the ocean. Grissom's comments after the flight reinforced suspicions that he had panicked and blown the hatch open himself. Immediately after the flight, a reporter asked Grissom whether he felt in danger during the flight. Grissom admitted being "scared a good portion of the time." In disbelief, a reporter asked: "You were what?" "Scared!" Grissom retorted. "Okay?"[62]

Grissom tried to restore his tarnished image after the flight. In an article in Life entitled "Hero Admits He Was Scared," Grissom said, "I was scared and I meant it." One would have to be abnormal not to be a "little frightened" by space flight, he continued, but he insisted that "fear never got the better of me."[63] In all of his comments after the flight, Grissom continued to insist that he had not panicked and blown the hatch himself: "I was just lying there minding my own business when the hatch blew."[64] Nonetheless, the damage had been done to his image as a great frontier adventurer. Whatever the truth, the sinking of the spacecraft could hardly be "sold" as a glorious adventure. Grissom, it seemed, simply lacked the "right stuff."

In another sense, however, Grissom's failure may have worked to NASA's advantage. The nation would no longer view the Mercury missions as merely short, safe rides. Grissom's near-tragic accident brought the important element of danger to the flights, adding a greater degree of credibility to the administration's depiction of further missions as frontier adventures. Grissom's personality and actions, moreover, added credibility to the administration's assertions that the astronaut's character and courage could influence the success of the flights.

Following Grissom's abysmal performance, the administration's frontier adventure became a story in search of a believable leading man, a more exciting plot, and a happier ending. In August 1961, the Soviets did their part to inject some excitement into the plot by sending Gherman Titov on seventeen orbits of the earth. Grasping the significance for the world of Shepard's "control" of his craft, the Soviets countered by portraying Titov as manually controlling his capsule too.[65] The malevolent antagonist had struck again. Now it was NASA's turn. The agency immediately canceled its one remaining suborbital flight and announced that the next flight would orbit the earth. On 28 November 1961, NASA took one more cautious step, sending a chimpanzee named Enos successfully into a dual orbit. But soon thereafter, the agency announced that the third manned flight would be a triple orbit of the earth lasting more than four hours.

After canceling ten scheduled launches over a two-month period, NASA finally sent its first manned orbital flight into space on 20 February 1962. Millions of Americans watched John Glenn blast off. From the start, Glenn's flight had the makings of a highly dramatic frontier adventure. On the second orbit, NASA received a signal from Glenn's craft that a heat shield was malfunctioning. Without the protection of the shield during reentry, the craft might disintegrate. Later, Glenn reported seeing "thousands" of tiny "fireflies" that glowed in the black sky. Attesting to the unknown element of the space environment, Glenn announced that "the true identity of these particles is still a mystery."[66] Ultimately, Glenn made a fiery reentry, watching "flaming chunks" six to eight inches long fly by his window. For a short period, NASA lost radio contact with Glenn. "It left me," Glenn wrote, "alone with my problems."[67] Robert Voas, Project Mercury's training officer, characterized Glenn's reentry as a supreme test of the heroic individual: "Glenn faced his moment of truth inside a fireball."[68] But when it was all over, the story had a happy ending. Glenn landed safely from his harrowing trip, and instantly he became a national hero.

Not since Charles Lindbergh's completion of the first transatlantic solo flight had the country expressed so much adulation of an American "adventurer." Glenn's flight clearly evoked the spirit of a frontier adventure. Unlike the suborbital flights, his triple orbit was spectacular enough to seem a real adventure. Glenn also emphasized that his flight would "pave the way" for voyages to the "moon and beyond"[69] and thereby fueled expectations of even greater adventures still to come. The mission's ten delays even contributed to the buildup of the flight and its "pilot." NASA had disclosed Glenn's identity immediately after the flight of Enos the chimpanzee, "a considerable change from the tighter news policy regarding crew selection" in the past.[70] As Americans became more familiar with Glenn with each postponement, anticipation and anxiety over the mission continued to build.

Most significantly, Americans could easily see that Glenn fit the pioneer role; he was obviously ready to play the part. Unlike Shepard and Grissom, Glenn did not contradict the administration's frontier mythology; he willingly contributed to the image-making. *Life* reporter John Dille wrote that more than any of his colleagues, Glenn had "the most consciously thought-out image of what an 'Astronaut' should be and how he should behave," both publicly and privately.[71] According to Dille, Glenn saw himself as "the first of a new and even heroic breed of men who have the enormous responsibility of serving as symbols of the nation's future."[72]

Glenn seemed naturally to possess all of the Puritan traits of the traditional heroic adventurer. Americans learned that Glenn cherished traditional Puritan virtues. A deeply religious man, Glenn proposed he had made his "peace" with his "Maker" years ago. His was not a "'fire engine' type of religion," something he called on only in emergencies and then put "back in the woodwork."[73] President Kennedy helped Americans interpret Glenn's statements about his faith, saying they reflected "a quality which we like to believe and I think we can believe is much a part of our American heritage."[74] In addition, Glenn demonstrated his devotion to family, breaking protocol during his speech before a joint session of Congress to introduce members of his family sitting in the audience, especially the "real rock" in his family, his wife, Annie. During training, Glen opted to forgo the sexual companionship and the comforts of living at home. Instead he stayed in the bachelor's quarters on base, running two miles every morning and, of course, attending church every Sunday. Glenn's self-control and sacrifice evoked memories of early Americans.

Coupled with the Puritan qualities, Glenn demonstrated qualities reminiscent of the early pioneers. Glenn, not "physically afraid," proposed that if an astronaut was "so shook up" he had to stay busy to remain calm, then he did not belong in the program.[75] In a postflight interview, Glenn described his feelings upon reentry as "cautious apprehension." Glenn reported he had "some concern" when he learned about his possible emergency.[76] The word "scared," however, never entered his vocabulary. On the contrary, Glenn called weightlessness "pleasant." He added that one could become "addicted" to space flight "rather rapidly."[77] Emotionally, Glenn seemed to have proven himself of good pioneer stock. During the long delays and possible emergency, he remained unruffled. He seemed to become emotional only when he spoke of his patriotism: "I still get a hard-to-define feeling inside when the flag goes by—and I know that all of you do, too."[78]

Despite all of his accomplishments, Glenn remained modest. Although he felt "proud" of his trip, he "also felt a certain humility."[79] Clearly, Glenn had not fallen prey to the sin of hubris—that excess of pride that often plagued other "heroes." Thus, when the Speaker of the House invited Glenn to speak before a joint session of Congress—an honor usually reserved for heads of state—Glenn stressed the accomplishment of the group over his own as an individual. Glenn acknowledged the great "honor" being "shown us."[80] My flight, Glenn pointed out, "involved much more than one man in the spacecraft in orbit."[81]

While Glenn demonstrated humility with talk of the team effort, he never forgot the importance of stressing the need for "rugged individuals" to "control" the capsule in space. Before the flight, Glenn commented, "I know at some point during the mission I will feel very much on my own."[82] After the flight, he proclaimed that the mission would have failed without a man "aboard to assume control and bring the capsule back."[83] "Flying" the capsule "myself," Glenn argued, proved "man's capabilities are needed in space."[84] As in Shepard's testimony, Glenn tried to characterize himself as a pilot, not a passenger. Never considering a Mercury astronaut as "merely a passive passenger," he stated that his flight had proven that man plays a "key role."[85] In the future, Glenn predicted, we can "put less automation into the machines" and rely even more on man by making him "a part of the system."[86]

NASA officially echoed Glenn's remarks about the need for brave "pioneers" in the space program. Suggesting that Glenn had succeeded through skill and daring, NASA doctor Stanley White proposed that Glenn's "one big task" during the flight was "control" and that his "manual flying" had exceeded the automatic flying.[87] D. Brainerd Holmes, director, Office of Manned Space Flight, similarly characterized Glenn's control of the craft as crucial to the mission's success. After discovering the malfunction, Glenn "assumed manual control of the spacecraft," Holmes explained. Without Glenn, added Holmes, NASA probably could not have accomplished the "full three orbits."[88]

Out of the public spotlight, both astronauts and NASA officials often admitted that the astronauts did not actually "fly" the Mercury capsules at all. As a star witness before the House Committee on Science and Astronautics, Glenn acknowledged that man did not operate in space flight as he did in airplanes.[89] Indeed, Glenn asked for changes in spacecraft design that would remove a great deal of the automation and place more reliance on man. He described the existing Mercury design concept as one capable of "doing every action that you want done in space and using man as a passive passenger to back it up, a man who only gets called up in case he is needed when a system malfunctions."[90] Asked whether people are under a mistaken impression that astronauts can control where they go in their crafts, Glenn admitted that the Mercury astronauts could control the craft's "attitude" but cannot "actually alter its flight path." We are looking forward to the time, Glenn added, "when man will really take his real, rightful place in space."[91]

The testimony of DeMarquis D. Wyatt, director, Office of Programs, before the House Committee on Appropriations was equally candid. Glenn, observed Wyatt, "did not in fact navigate or guide. All he did was control the stabilization of the spacecraft." In future spacecrafts, Wyatt added, astronauts will "actually fly" the craft in addition to stabilizing it. He went on to describe Mercury flights as "purely mechanical systems."[92]

In the public mind, there seemed little doubt that Glenn's flight was a great adventure on the New Frontier. Glenn, now the model of the modern frontiersman, had battled the elements, had reacted to setbacks with courage, and had returned unscathed. Unlike Shepard and Grissom, Glenn cherished the image as a heroic adventurer and willingly contributed to the image. Glenn's flight was worthy of a frontier adventure, and NASA publicized the flight and its astronaut prior to the launch, allowing Americans to become familiar with Glenn and to develop needed anticipation and concern for the flight and its "pilot." Before and during the flight, John Dille explained, Glenn "portrayed the perfect image of the modest, dedicated and patriotic hero." He probably did more on 20 February 1962, said Dille, "than dozens of others could have done in months to sell the U.S. space effort to Congress and to the nation."[93] Walter McDougall termed America's outpouring of emotion for Glenn's feat "a national catharsis unparalleled in the quarter century of the Space Age."[94] Glenn brought believability to the story, but the effect of the flight was not permanent, as subsequent events would prove.

Three days after Glenn's flight, President Kennedy anticipated a major theme of frontier mythology when he spoke of the possible temptations and pitfalls awaiting America's new heroes. The astronauts, he said, will learn that "the hazards of space flight only begin when the trip is over."[95] Kennedy knew that Americans would watch their heroes carefully. Americans knew the story almost instinctively; temptations were always greatest at the top.

Two events in mid-1962 suggested that the astronauts may indeed have succumbed to worldly temptations from an excess of pride. First, questions arose about a contract that the astronauts had signed with *Life* for their stories. Many viewed the deal as unethical, since the astronauts, employees of the government, benefited financially from their assigned duties. Second, the public learned that the astronauts had agreed to accept free homes from a Dallas development group. NASA had to hold a news conference to explain the actions of the astronauts and their attorney. In the wake of substantial negative publicity, the astronauts reversed their position and declined the homes.

Meanwhile, the flights continued, with Scott Carpenter, Wally Schirra, and Gordon Cooper all piloting Mercury missions within the next year. Following "the trail blazed" by Glenn,[96] Carpenter blasted off on 15 May 1963. Like Glenn, he had difficulty during reentry. Even though his fuel became "dangerously low," he did not panic, enabling him to bring his craft back safely. Schirra, the third American to orbit the earth, attempted to characterize his flight as a "real breakthrough in manned space flight." Schirra asserted that upon liftoff he had turned off all automatic sequences: "The capsule was all mine now."[97] In the past, continued Schirra, ground stations had kept tight control of the situation. Now they trusted me with "the works." Seeking to magnify the significance of his flight, Schirra declared that nobody had ever "flown a capsule before, much less under full pilot control."[98]

Although both Carpenter's flight and Schirra's flight were successful—even groundbreaking in some respects—enthusiasm for the "adventure" seemed to wane in late 1962. Despite NASA's best efforts to emphasize the importance of man's role in the program, serious questions about the need for a *manned* space program began to emerge.[99] Glenn's flight and the successful confrontation with Cuba in October 1962 restored the nation's pride, thus eliminating the urgency behind American support for manned space flight and leaving administration officials to take additional measure in 1963 to retain support for Project Apollo.

With the coming of 1963, NASA began preparing to sell its budget to an increasingly critical Congress. It was in this context that Gordon Cooper's flight—the last Mercury mission—took on additional importance. Concerned with growing criticism, the Kennedy administration went to great lengths to recapture the glorious "frontier" excitement surrounding the flight of John Glenn. NASA conducted an enormous public relations campaign to engender excitement and support for Cooper's flight. The press kit for Cooper's Mercury-Atlas 9 began like most of the others, referring to Cooper "at the controls." Yet it emphasized a transformation in the astronauts. "The astronauts have changed through their flight training and actual flight experiences, proving themselves space pilots rather than spacecraft passengers."[100]

The flight itself seemed worthy of the big buildup. Cooper completed twenty-two orbits of the earth, staying in space for over twenty-four hours. During the flight, Cooper lost all automatic controls, and like a true pioneer, he "flew" his capsule back to earth. During his flight, Cooper explained, his automatic control system malfunctioned and he had to assume control. Positioning his craft for reentry and riding the capsule back from space "was going to be up to me."[101] Cooper returned to a hero's welcome.

In perhaps NASA's best public relations move yet, the agency scheduled Cooper's flight to coincide with the anniversary of Lindbergh's solo flight across the Atlantic. Further dignifying the occasion, President Kennedy presented NASA's Distinguished Service Medal to Cooper in a White House ceremony.[102] In remarks during the ceremony, Kennedy compared Cooper's flight to Lindbergh's achievement, calling them "equally hazardous" and "equally daring."[103] Next, the Speaker of the House invited Cooper to address a joint session of Congress. The nationally televised speech gained further exposure for Cooper, NASA, and Project Apollo.

Thus concluded what President Kennedy called "an extraordinary page in American history." Kennedy praised the Mercury astronauts, who had "become part of the American story in a very real way."[104] By sending a man to the moon, Kennedy observed, Americans would assure themselves of "playing their great role, as they have in the past."[105] At the conclusion of the Mercury program, Americans finally seemed firmly committed to continue "playing their great role" in space exploration.

Although manned space exploration did not inherently possess all the constituents of a traditional frontier adventure story, the American people could see enough parallels to conjure up the powerful frontier mythology. The moon presented a tangible, conquerable "scene." To reach it, the astronauts had to journey through the unknown environment of space and overcome a human villain, the Soviets. The astronauts reacted intelligently in dangerous situations, "flying" their crafts to safety. With each successive mission, the astronauts demonstrated greater and greater control of their crafts, proving the importance of individual "pioneers" to the program.

Various obstacles impinged upon the administration's ability to peddle its first two flights as frontier adventures. Alan Shepard's and Gus Grissom's comments shunning the hero image, the unspectacular nature of their flights, the administration's failure to engage in preflight publicity, and the sinking of Grissom's capsule all made the comparison too unbelievable to sell. With John Glenn's successful orbital flight, the administration evoked the frontier mythology and captured the imagination of the American people. Glenn seemed naturally to fit the mold of the heroic adventurer. But the public's adulation of Glenn did not last indefinitely. After Glenn's flight, excitement for the manned moon mission began to wane. Gordon Cooper's flight, however, recaptured some of the previous excitement, helping the administration quiet serious criticisms of manned space exploration.

Fittingly, the nine new Apollo astronauts, introduced in late 1963, took over where the original seven astronauts had left off, adopting the romantic frontier rhetoric of their Mercury brethren. In a *Life* article cowritten by the new astronauts, Astronaut Elliot See underlined the frontier narrative when describing how his attitude about the moon shot had changed during his Apollo training. "The whole mission," he proposed, "becomes more real to me and less of an adventure into the unknown."[106] The first man on the moon, Neil Armstrong, also supported the administration's view of the manned lunar landing. Initially, he insisted, he did not volunteer for Project Apollo because he was "skeptical" of the program. The success of the flights, however, made him change his mind and realize that he had "underestimated" the astronauts. Armstrong also attributed his initial reluctance to join the space program to a lack of a clear objective. "Another thing that affected my thinking was that there was no moon program in those early days."[107] Still another astronaut, Edward White, who later died tragically in the Apollo fire, adopted the language of an explorer. Calling people of Earth "pretty curious" about the moon's composition, he maintained that man would never "satisfy" his "curiosity" unless man himself went to the moon. Like the Mercury astronauts before him, White stressed the most crucial element of the frontier motif: astronaut control. "The most important thing," White contended, "is that man—not the automatic machine—is the primary system in space flight."[108]

Even with the Mercury and Apollo astronauts echoing the administration's depiction of a manned lunar landing, the group could not sell the moon shot by themselves. The administration had limited opportunities to speak directly to the public. Moreover, most Americans would learn of the space program from the media.

3

Media Coverage of the Space Program: A Reflection of Values

In assessing the press coverage of American space exploration in its 8 October 1962 issue, *Newsweek* calls the men assigned to interpret the space age "a new kind of journalist," trained largely in the post-Sputnik era.[1] The amount and complexity of the scientific information they had to master, the "wilting ignorance" of their editors about science, and the roadblocks NASA erected to stop the press from getting information reflecting badly upon the program made the conditions under which these reporters operated "unique."[2] The writers, moreover, did not even agree on their function. Should they educate the public in "the intricacies" of the space age or simply report events? Should they serve an adversarial role, or should they celebrate the space program as a great national endeavor? *Newsweek* acknowledged the latter dilemma in the minds of many reporters. As citizens of "a nation whose prestige is at stake in the space competition," *Newsweek* comments, "the reporters are under the temptation to function as rooters for 'The Team'—a role abhorrent to most newsmen."[3]

Researchers have focused primarily on the press coverage of the early space program and have neglected television coverage for a number of reasons. First, although television undoubtedly played an important role in the impression Americans formed of the space flights, television was still in its infancy in the early 1960s. Not until 1963 did network evening newscasts expand from fifteen to thirty minutes. Thus at the height of the program in 1961 and 1962, the networks had little time to devote to America's space efforts. Second, color television was still a thing of the future in the early 1960s. Finally, the central problem with studying television news coverage of the early space program is the lack of videotape in archives.

Scholars who have studied the press coverage of America's early space program disagree about whether the press was biased in favor of the administration's lunar landing project. Robert Cirino attacks the coverage of the manned space program, calling the media "the willing partner of the NASA propaganda machine."[4] Specifically, Cirino charges that NASA and the press gave the public the impression that favoring the space program meant favoring a moon shot, while opposing the moon landing meant opposing the entire space program. The press "intentionally" failed to inform the public that most of those who opposed manned space exploration favored unmanned space exploration. The news media, according to Cirino, concealed the opposition to the lunar program by ignoring it.

Ronald E. Ostman and William A. Babcock, on the other hand, argue that the media exhibited no "pro-Kennedy bias" in reporting on the space program.[5] The authors examined three major newspapers' coverage of the manned space program and found that the vast majority of the articles in the papers presented neutral stories. They identify a handful of articles with detectable biases. The "biased" articles, however, according to Ostman and Babcock, were almost equally split between those for and against Kennedy's space program.

Thus questions about the role of the media in building support for the Kennedy administration's space program remain. As Ostman and Babcock point out, Cirino "generalized far beyond the data he presented."[6] Although Cirino asserts that one can find media bias in favor of manned space flight from the beginning of the space program, he offers little evidence to support his contention.[7] But Ostman and Babcock themselves examine only three newspapers, and they altogether ignore newsmagazines and other media. They, too, offer only a limited view of news coverage of the space program in the Kennedy years.[8] More important, their analysis fails to address Cirino's contention that the press was biased toward *manned* flight and ignored calls for unmanned space exploration. Ostman and Babcock ask merely whether one can find "a perceptible bias in newspaper reporting of the U.S. space technology and exploration issue."[9] They simply overlook the distinction between manned and unmanned exploration.[10]

The question of bias in popular press coverage of America's early manned space program is more complex than described by either

Cirino or Ostman and Babcock. In examining all articles on the space program in the *New York Times* and in the thirty best-selling magazines in America from 1959 to 1963, one finds a variety of critical and uncritical coverage that defies simple categorization as for or against manned space exploration. Certain magazines provided little negative coverage throughout the space program. Rarely does one find an article critical of the manned space effort in *Life, Popular Science,* or *Popular Mechanics.*[11] Overall, *Time's* coverage supported a manned lunar landing.[12] Yet events in late 1962 and early 1963, more than a change in editorial policy, prompted *Time, Newsweek, Reader's Digest,* the *Saturday Evening Review,* and the *New York Times* to criticize certain aspects of the administration's program to put a man on the moon. Criticism of the moon shot by respected scientists, the astronauts' signing of a second contract with *Life* for their personal stories, the Government Accounting Office's findings of expensive mismanagement in NASA, NASA's own study of shoddy workmanship by contractors, continuing charges of pork-barrel politics, the president's call for a joint U.S.-U.S.S.R. space effort, the Soviets' apparent withdrawal from the lunar race, and the lack of an American manned space flight after May 1963 all contributed to the press's more critical posture.

From the beginning of the manned space program in 1959, the *New York Times* offered frequent coverage of America's space program. Occasionally, the paper described the minimal "control" the astronauts exerted during their missions. Rarely did it challenge the astronaut's assertions that they "flew" or "controlled" their crafts during their missions, even though its news stories at other times explained that the astronauts could merely "control" the stabilization of their capsules.[13] Contrary to Cirino's charge, the paper often mentioned unmanned exploration of the moon as an alternative to manned flight.[14] From NASA's inception, the newspaper's editorials and columnists supported the manned space program. By 1963, however, one finds a marked change. The paper began calling for a reexamination of the moon program and its emphasis on manned flight.[15] Interestingly, in the midst of the paper's critical posture in 1963, Gordon Cooper's flight received positive coverage, demonstrating that the drama of manned flights still evoked patriotic pride.

The *Saturday Evening Post, Reader's Digest,* and *Newsweek* offered the most critical coverage of the space program. These magazines frequently attacked the military, political, scientific, economic, and technical value of sending a man to the moon. Again, contrary to Cirino's assertions, *Newsweek* discussed unmanned lunar shots as an alternative to manned flights.[16] *Newsweek,* like the

New York Times, showed two faces. Although it attacked Project Apollo and explained the limited role that the astronauts played during their missions, the publication abandoned its critical posture when describing the manned flights. In sum, when examining the coverage of America's manned space program in the popular press from 1959 to 1963, one does find support for manned flight. The press does not, as Cirino charges, ignore unmanned exploration. The press offered both positive and negative coverage.

Thus a simple charge of bias for or against manned space flight does not adequately explain press coverage of America's efforts to place a man on the moon. Rather, the mixture of critical and celebratory coverage reflected certain enduring values in journalism. These enduring values, moreover, predisposed journalists to cover the space program in a way that supported, even glorified, a *manned* space program.

Negative Coverage

Herbert J. Gans proposes that journalism does not confine itself to reality judgments but also "contains value, or preference statements," that underline the news and present a "picture of a nation and society as it ought to be." These implicit values, he adds, often "affect what events become news and even help define the news."[17] One of the most prominent values, according to Gans, dates to Thomas Jefferson's celebration of small-town America. This preference for small-town pastoralism translates into a more general value of the desirability of smallness. One can see bias toward smallness in stories that examine the faults of largeness. "In the news," Gans argues, "big business, big labor, and big government rarely have virtues."[18] Thus, when reporting on the size of NASA, particularly its rapid growth, *Time* lamented that America's space program had "sprouted like Jack's beanstalk, sucking up men and money at a prodigious rate, sending its tendrils into every state."[19] Similarly, Stuart H. Loory emphasized the perils of the program's rapid growth in a 14 September 1963 article in the *Saturday Evening Post*: "Big, blaring, burgeoning in a hundred directions, the space program stands accused today as a monstrous boondoggle."[20] Much like muckrakers at the turn of the century, the media during the early 1960s decried the growth of the space program by comparing it metaphorically to a monster. As *Time* put it on 4 October 1963, "Infant space industries [have grown] overnight to monster maturity."[21]

In the enduring value system of American media, bigness goes hand in hand with waste. Because the program had "grown too rapidly," reported John Finney in the *New York Times*, "waste and duplication" had become commonplace.[22] Stuart Loory reflected the same value in the *Saturday Evening Post* when he attacked NASA's "expanding bureaucracy" for creating "confusion and inefficiencies."[23] Other articles in the *New York Times* went even further. An editorial published on 28 June 1963 not only attacked NASA's "waste and duplication" but attributed it to the House Science and Astronautics Committee's misguided fondness for all things large. "The committee's largesse and laxity," the paper stated, "encouraged NASA to act as if there was no limit to what it could spend in the skies."[24] As one might expect, the newspaper's solution to the problem of waste was simply to reduce NASA's size. Cuts in NASA's budget, wrote John Finney, would eliminate "waste and duplication" and would tighten NASA's management.[25] In short, bigness encouraged waste, and smallness encouraged efficiency. The media had no bias against the space program generally but simply criticized it on occasions when its size seemed antithetical to the values of small-town pastoralism.

The majority of the media's negative coverage of the Kennedy administration's space program focused on alleged self-interest and partisanship infesting the program. Again, one can best explain the coverage as a reflection not of an anti-Kennedy sentiment but of an enduring journalistic value. American news indicates how American democracy should perform by frequently reporting on deviations from an unstated ideal. One may label this ideal altruistic democracy. As Gans observes, "the news implies that politics should be based on the public interest and service."[26] Thus nepotism, logrolling, patronage appointments, financial corruption, and anything generally viewed as a "deal" is always news. Reflecting this value, journalists inevitably criticize any decision "based, or thought to be based, on either self-interest or partisan concerns."[27] In covering the space program, the press again found evidence of wasteful spending conflicting with this value of altruistic democracy. In a number of stories, Congress became the target for allegedly failing to examine the space budget critically.[28] The press charged that until 1963, Congress served as the space program's "sugar daddy," giving it virtually everything it demanded.[29] One can also see this value reflected in stories criticizing administration officials, and even the astronauts themselves, for placing personal rewards above public service. Starting in late 1962, the media raised two specific issues to the top of its space coverage: the astronauts' renewal of a contract for their personal stories and the administration's favoritism in the selection of sites for space facilities.

The astronauts' contract to sell their "personal stories" received widespread coverage in the press. The terms of the contract itself, distinguishing the astronauts' personal stories from their public stories as government employees, made it newsworthy. Although it reported the signing of the astronauts' first contract with *Life* in 1959, the press did not attack it. With the signing of the second contract in late 1962, however, coverage increased and criticism surfaced. The *New York Times* ran numerous editorials blasting the contract. One editorial, for example, proclaimed that the government should not allow the astronauts "to reap enormous private profits" from participating in "a great national effort."[30] News stories in the paper also emphasized that the astronauts' personal gain came at the taxpayers' expense. The government followed an "inappropriate" policy, the *New York Times* concluded, when it allowed the astronauts, whose stories belonged in the "public domain," to collect money from "a private payroll."[31]

Not surprisingly, the two leading newsmagazines differed in their view of the contract between NASA and *Life*. Although *Time*, *Life*'s sister publication, did not seriously criticize the contract, its major competitor did. "How much a hero can expect to gain financially and still remain a hero," *Newsweek* observed, "is uncertain."[32] *Newsweek* attacked the contract as an "embarrassing financial arrangement" and belittled contract negotiations as "legal bickering" more appropriate to the film *Cleopatra* than to a "serious scientific endeavor."[33] The magazine also blasted *Life*'s portrayal of the astronauts. The program, *Newsweek* declared in its 12 February 1962 issue, had begun to resemble a "Barnumesque extravaganza" and the astronauts, the "cardboard characters of soap operas."[34]

The media focused even more attention on the alleged favoritism of NASA and the administration in awarding lucrative space contracts. One finds the value of altruistic democracy in the media's attack of the Kennedy administration and Congress for the "aroma of pork-barrel" they allowed "to spring up" around the program.[35] Reporters exposed and then attacked partnerships between administration officials and members of Congress. The press questioned, for example, the "coincidence" of NASA's placement of its Manned Spaceflight Center in Houston, the home district of Democratic Representative Albert Thomas, chairman of the House appropriations subcommittee. Not surprisingly, wrote *Newsweek*, "Thomas wasn't very sympathetic to space spending until the question of a new astronaut center came up."[36] It did not help matters when Vice President Lyndon B. Johnson, who had previously told reporters Texas would get its fair share of the space contracts, announced the award of the spaceflight center from his Houston office.[37]

Two individuals attracted a great deal of the media criticism: Senator Robert S. Kerr of Oklahoma, chairman of the Senate Aeronautical and Space Sciences Committee, and NASA administrator James E. Webb. Before assuming his post as administrator, Webb served as the assistant to the president of Kerr-McGee Oil Industries, Kerr's own company. *Time* reported that critics had attacked Webb for awarding space contracts with "a political rather than a scientific eye."[38] The Webb-Kerr partnership, reporters charged, amounted to patronage. Although all members of Congress should be equal in the fight to land "juicy space contracts for their home states," *Newsweek* observed, some members are "more equal than others." The magazine added that because of his connections, Kerr was "the most equal of all."[39]

The press waged its most savage attack on the president and his brother Senator Edward M. Kennedy for a $50 million electronics research facility that NASA awarded to Boston, Senator Edward Kennedy's home district. The award came only months after Senator Kennedy had successfully campaigned on the pledge that he could "do more" for Massachusetts. What made the facility doubly suspicious, according to media reports, was that NASA broke with its usual practice of establishing formal criteria for the need and location of the site. NASA also failed to have a board review the location. Even more suspicious, the project became a last-minute addition to a budget previously reviewed by the Bureau of Budget.[40]

A *New York Times* editorial on 3 August 1963 suggested that the president himself was probably behind the decision. Observing that Kennedy had kept his hot line to Congress "sizzling" in the past few days with pleas to approve the Boston facility, the editorial explained that the Senate had reversed its earlier decision to reject the facility. "Of course," the paper sarcastically added, "no one would be cynical enough to believe that the calls could have anything to do with the decision."[41]

Positive Coverage

The vast majority of the coverage of the space program was positive. Like the negative coverage, the celebratory coverage did not simply reflect a partisan bias. Instead, certain aspects of the space program appealed in a positive way to enduring media values. "One of the most important enduring news values," Gans proposes, "is the preservation of the freedom of the individual against the encroachment of nation and society."[42] W. Lance Bennett echoes Gans when discussing the journalistic imperative to personalize news. Bennett defines personalized news as a "bias that gives prefer-ence to the individual actors and human-interest angles in events while down playing situational and political considerations that establish the social contexts for those events."[43]

The media's ideal individuals successfully struggle against adver-sity, overcoming forces more powerful than themselves. The news particularly seems to celebrate individuals who "conquer nature" without harming it: explorers, mountain climbers, and of course astronauts. With the majority of the news about the space program focusing on the flights themselves, the media's model of the ideal individual predisposed reporters to write of the individual heroically struggling to overcome unknown, powerful forces.

The media's ideal of individualism is most evident in the coverage of the flights of John Glenn, Walter Schirra, and Gordon Cooper. Suggesting that the flights were individual rather than team tri-umphs, *Newsweek* proposed that the greatest lesson one learned from Glenn's flight came on a subjective level. "The drama of the human spirit—solitary, vulnerable, curious—facing the unknown elements of the universe is as old as mankind."[44] Glenn, the article adds, demonstrated that Americans still exist who can play the heroic role in this "ageless drama." Notice the similarity of *News-week's* description of Gordon Cooper's flight in 1963: "Once more, the ancient drama of the solitary individual against the elements was re-enacted."[45]

Examination of the press coverage of the astronauts *before* the Kennedy administration came to power demonstrates that the me-dia's depictions of the astronauts and the program reflected not a pro-Kennedy bias but an enduring premium on individualism. In its first issue on the astronauts on 20 April 1959 *Time* described the astronauts in two separate articles as "individualists all."[46] One can find similar descriptions of the astronauts three years later when the press began emphasizing the astronauts' isolation, even loneliness. On 5 February 1962, for example, *Newsweek* called Glenn "a single remote figure" and compared him to Charles Lindbergh, another "authentic individualist."[47] Other articles described the astronaut as "alone" and "lonely." *New York Times* writer Richard Witkin, for example, described Gordon Cooper during his flight as "the pilot alone in orbit."[48] Individualism dictated the *New York Times'* de-piction of Gus Grissom's childhood. The paper reported that as a schoolboy, Grissom slipped off for "solitary swims" in quarries, explored limestone caves "alone," and made all-night camping trips "by himself."[49]

Other stories emphasized comparisons between the astronauts and the mythologized "individualists" of the past, including aviators Charles Lindbergh and Orville and Wilbur Wright and the New World explorers Columbus and Magellan. Occasionally the press compared the astronauts to both groups. Richard Witkin, for example, wrote that neither "Columbus' opening of the New World" nor "the Wright brothers' first flight had consequences as profound as may emerge from the first lunar voyage."[50] More often, the press, particularly the *New York Times*, compared the astronauts to explorers of the New World. An editorial on 11 April 1959 set the tone for future descriptions of the astronauts. The first man to fly into space, the *New York Times* asserted, would "assure himself immortal fame alongside Columbus and Magellan."[51] A *Time* article, written one week later, demonstrated that the editorial was not merely an isolated instance. The astronauts, the magazine proclaimed, were "cut from the same stone as Columbus" and "Magellan."[52] Years later one finds the media using this comparison, focusing attention on the efforts of a heroic individual in describing the flights or interpreting their meaning. Perhaps the best example comes from *New York Times* columnist James Reston. After Scott Carpenter's flight in May 1962, Reston, in his usual postflight hysteria, asserted that the astronauts "may make Columbus and Vasco de Gama look like shutins before they are through, and their exploration may open up more in the heavens than the old sailors did on the sea."[53] Reston, moreover, made a similar comparison when first describing the astronauts on 12 April 1959—nearly two full years before Kennedy came to office.[54]

The press' focus on individualism and its comparisons between manned space flights and the exploits of Columbus, Magellan, and the Wright brothers led the media naturally to embrace the space program as a great frontier adventure story. This traditionally American story, with its emphasis on rugged individualists, fit perfectly within the media's celebration of the ideal individualist. In doing so, of course, the media uncritically adopted the perspective of the Kennedy administration's public relations campaign and, in effect, argued against critics of manned space flight or of the scope of America's commitment to space generally. From the start, moreover, the media demonstrated a predisposition toward describing the space program as a frontier adventure. A *New York Times* editorial on 11 April 1959 spoke of the "extreme demands" the Mercury "adventure" would place on the "first pioneers" of space.[55] Early articles depicted the astronauts as part of a special breed destined for greatness. A *Newsweek* article dated 20 April 1959 proposed that the seven astronauts "bore a special stamp that set them apart."[56] *Time*'s 20 April 1959 issue asserted that the "curious finger of fate" had selected the astronauts to be "hurled into space to make the supreme test."[57] In a different article in the same issue of *Time*, the magazine placed the astronauts within a long line of heroic pioneers: "Rarely were history's explorers and discoverers so clearly marked in advance as men of destiny."[58] Although one can see the brief outlines of the frontier motif in a few articles in 1959 and 1960, tremendous emphasis did not fall on the story until Kennedy came to power.

Couching the story of the space program in terms of "the new frontier," Kenneth Crawford of *Newsweek* stressed how *all* Americans were the "true heirs of a frontier tradition." Crawford placed the program above political disagreements, insisting that Americans *take for granted* that they "shall be the pioneers who take advantage of the opening."[59] Similarly, *New York Times* Washington correspondent John Finney stressed the virtual inevitability of Americans' exploration of space in terms of frontier mythology: "Just as in ages past, the first explorers can be expected to be followed by the settlers and the military along the new frontier."[60]

Even if the administration had not portrayed manned space exploration as a frontier adventure story, the media might have done so. The story offered two attractions to American journalists. First, it offered conflict, a treasured element in journalism. Using the frontier story to describe the space effort allowed journalists to pit American astronauts not only against Soviet cosmonauts but also against the unknown environment of outer space. Second, the frontier story appealed to one of the most deeply rooted journalistic values: rugged individualism. More than any other story of the times, the space program appealed to the media's interest in stories of individual success, made possible by the old-fashioned virtues of the frontier adventure.[61]

The astronauts themselves provided the stories for the media's version of the "frontier adventure" in space, reflecting the journalistic imperative to personalize news.[62] Rather than the technicians and scientists who made it all possible, the astronaut as frontiersman came to represent the entire program. Thus a *New York Times* editorial in March 1962 credited the astronauts, not the technicians, with being "the young pioneers of the space frontier."[63] *Newsweek* likewise credited the success of the program largely to the fact that the astronauts had a "frontiersman's drive to stake out new territory."[64] Even *Popular Mechanics*, in its March 1959 issue, seemed more interested in the astronaut as a "Daniel Boone of space" than in the technological heroes of the program.[65] In virtually all coverage of the space program, it was the astronauts rather than the scientists who were blazing trails for others to follow.[66]

It mattered little which of the astronauts was the center of attention. A *New York Times* editorial labeled Scott Carpenter "in the tradition of the pioneers of a century ago."[67] A year later, one finds very similar imaging in *Time*'s description of how Gordon Cooper executed a manual reentry. "Like a rifleman with a cross-hair sight," recounted the magazine, Cooper "lined up the horizontal mark on his window with the horizon."[68] The press even interpreted Cooper's speech following the flight as something from the frontier era, with *Newsweek* praising the "homespun words" of the address.[69]

Clearly, however, one astronaut stood out in the press coverage as a man who best exemplified the whole range of traditional American values associated with both the Puritans and the pioneers. Because he so completely exhibited all the personal qualities that made the space program successful, John Glenn became the quintessential American astronaut.

Media enthusiasm for Glenn went beyond admiration for his flight. The press may well have celebrated anyone who put America back in the space race. "The surprise," reported *Time*, "was that [America] found Glenn the man fully the equal of Glenn the astronaut."[70] Columnist James Reston agreed: "Glenn himself, is almost as important as his space flight, for he dramatized before the eyes of the nation the noblest qualities of the human spirit."[71] According to *Time*, "Glenn's modesty, his cool performance, his dignity, his witticisms, his simplicity—all caught the national imagination."[72] To a great extent, the media's ideal of individualism accounts for the emphasis on Glenn's personal qualities. Comparing Glenn with heroic Americans of the past also added credibility to the idea that his personal qualities and character accounted for the success of the entire space program. Placing Glenn in a long line of rugged individualists, the press suggested that success in space was as much a result of individual effort as the opening of the western frontier by individual pioneers.

Readers learned that Glenn exemplified the qualities and character of Americans of yesteryear. He possessed the self-reliance, courage, and patriotism of the pioneers who settled the frontier. *Newsweek* called him a "self-reliant, modest, and courageous man."[73] The *New York Times* particularly highlighted Glenn's courage. Arthur Krock deemed him "fabulously courageous," while James Reston asserted that the astronaut "dramatized courage."[74]

Glenn, as described by the press, also embodied the simple character of the American Puritan. *Newsweek* called him an "island of disciplined calm."[75] The stoic astronaut demonstrated his self-control when it counted: during his flight. Receiving news that his craft might disintegrate upon reentry, according to *Time*, Glenn reacted with "characteristic calmness."[76] By controlling his emotions, Glenn demonstrated his tremendous discipline. Glenn leads a "life of austere discipline," *Newsweek* observed.[77] Even his dedication to his physical training set him apart. The *New York Times* said running on the beach was a "routine followed daily by Colonel Glenn and less frequently by some of his six astronaut colleagues."[78] From the descriptions, one detects almost a masochistic tendency, also reminiscent of the Puritans.

The press reported on a variety of other qualities that also placed Glenn in the tradition of the Puritans. *Time* remarked that "through the whole ordeal of instant heroism, he continued to display a remarkable modesty and control."[79] His modesty and humility stemmed from his uncomplicated, simple nature that, according to *Time*, caught "the national imagination."[80] Even Glenn's family could not escape writers' oversimplifications. The Glenn family, wrote Gay Talese in the *New York Times*, possessed the "simplicity" of a "Norman Rockwell" original.[81] Fueled by his "determination," Glenn's dedication "stood out" even among his fellow astronauts. Glenn was the only astronaut, for example, who did not move his family to the training facility; he lived at the base, recalled *Time*, so that "he could better concentrate on the program."[82]

By far, the greatest amount of description, testimony, and commentary concerned Glenn's faith. To journalists, it must have been inconceivable that Glenn, facing the dangerous unknown all alone, would not have a deep faith to draw upon to sustain his courage and drive. One can see the press's assumptions at work in *Newsweek*'s first article on the astronauts. "Inevitably," the magazine declared on 20 April 1959, "the question of religious faith came up."[83] Numerous articles in the press described Glenn as "deeply religious," while one went so far as to refer to Glenn's room as his "monastic quarters."[84] The *New York Times* brought his family's faith into the picture, continually reminding readers that Glenn and his family attended church "every" Sunday.[85] In sum, to make Glenn conform to traditional views of heroic adventurers, the press focused attention on personal qualities reminiscent of the Puritans and pioneers. This comparison, moreover, became crucial in dictating the terms of the debate over manned versus unmanned space exploration.

Man Versus Machine

Traditional frontiersmen of the past conquered and dominated their respective frontiers by actively exerting control over them. Frontiersmen acted to tame their environments. The frontier narrative's focus on the individual, therefore, dictated portrayal of the astronauts, like pioneers of the past, as exerting control during their flight. They could not merely serve as passive passengers; they had to pilot their crafts. One can find isolated instances in which the press questioned the amount of control the astronauts actually exerted and even challenged the usefulness of manned flight. John Finney, for example, the Washington space correspondent for the *New York Times*, inadvertently questioned the control exerted by the Mercury astronauts when he described the difference between the Mercury and Gemini capsules. The Gemini capsule, he wrote, will be "under the control of the astronauts rather than automatic instruments. In effect, the astronauts will be flying the capsule."[86] His description contrasting the Mercury capsule with the Gemini capsule contradicted reports that the Mercury astronauts would actually "fly" their capsules. Richard Witkin more directly challenged the necessity of the astronauts after reporting in the *New York Times* that the Soviet Union had sent Valentina V. Tereshkova into space, a woman with no experience as a pilot. Witkin called America's emphasis on the need for professional test pilots "somewhat misleading."[87] Harland Manchester of *Reader's Digest* went even further, saying that the "encapsulated man" can do little "but go along for the ride and show whether he can take it."[88] Interestingly, in one article, *Time* questioned the amount of control space travelers exerted while nonetheless calling forth the frontier motif. "The earth-circling trips of the astronauts and cosmonauts," the magazine explained, "were almost as passive as floating down a river on an oarless raft."[89] Most of these descriptions, however, occurred during the calm between flights and paled in number by comparison with articles that focused attention on the individual piloting and control skills that the astronauts exerted during their heroic missions. At the time of the space shots, the press never seriously challenged the astronauts' assertions that they controlled their flights. The media's celebration of individualism actually seemed to lead reporters to search for evidence of "control" by the individual "pilot."

Thus *Time* commented that Alan Shepard's flight proved that the astronauts could "operate" the Mercury capsule, which the magazine described as far from "a passive" space vehicle "just up there to coast along."[90] The media also distinguished between Shepard's flight and Soviet Cosmonaut Yuri Gagarin's flight by focusing on the control each exerted. Gagarin served more as "a passenger than a pilot," the press observed, while Shepard took "control of his ship" and performed "five separate capsule maneuvers."[91] The press went even further in reporting the control that John Glenn exerted. According to John Finney, Glenn's flight "proved" that man "could and should be more than just a passive passenger aboard an automated spacecraft."[92] The media depicted Glenn as taking over complete control of his capsule, fulfilling man's ultimate role in the space adventure.[93] Similarly, the press focused on the individual effort of Gordon Cooper. In a breathless recounting of Cooper's flight, *Newsweek* proposed that when Cooper's automatic controls malfunctioned, he had to "pilot his spacecraft back from orbit by human skill alone."[94] John Finney asserted that Cooper's control of his craft played a part in the public's excitement over his flight. Cooper's use of the "manual controls to return his capsule to earth," Finney observed, "explains in large measure the hero's welcome he was accorded."[95] Robert Heinlein best summarized the view the press presented the public. "The Mercury shots," he proclaimed, "proved that an astronaut can actually control his ship."[96]

Obviously, the perception that the astronauts controlled their crafts was crucial to the coherence of the frontier narrative. The importance of the values clustered around individualism influenced the media's reporting of the space program, causing them to accept depictions of the astronauts as pilots who flew their spacecrafts. This, in part, explains the media's denigration of unmanned space efforts. As Gans explains, "the news often contains stories about new technology that endangers the individual" and deprives human beings of "control over their own lives."[97] Coverage of the space program provided a clear example of the media's tendency to celebrate man over machine, and this tendency in turn created a bias in favor of *manned* rather than unmanned space exploration.

The press took every opportunity to stress man's superiority to a machine. Man, the media asserted, had unique capabilities that machines could not match. Man could retrieve more information in a shorter time period. One finds an instance of this focus early in the space coverage. "Instruments," *Reader's Digest* proposed in April 1959, "can never bring back as much information as a spaceship with a human crew."[98] The media also claimed that a human observer could retrieve the increased amounts of information much more quickly than machines. Unmanned probes, Bernard Lovell argued, would take "decades" to find answers "trained geologists" could get "in a few hours on the lunar surface."[99] Ultimately, man's brain made him superior to machines. A *New York Times* editorial proclaimed that "no instrument on earth" could substitute for "a prepared mind" coming upon "unexpected observations."[100] Lovell pointed directly to the distinguishing characteristic between man and machine. Man must go to the moon, Lovell argued, because machines have "no brain."[101]

Besides its focus on man's unique ability to reason, the media also highlighted the individual by denigrating machines. As Gans remarks, the news "always" makes room for "gleeful" stories about machines breaking down. Undoubtedly, pointing out the shortcomings of machines helps individuals to maintain the perception that they retain control in an increasingly mechanized world. John Finney pointed to the American household to make his point. "As any housewife with an automatic washer can attest," Finney announced, "automatic equipment produced by an American industry attuned to mass, rather than quality, production can be exasperatingly unreliable."[102] The press particularly highlighted the failure of machines and the superiority of the individual in its coverage of the flights of John Glenn and Gordon Cooper. During Glenn's flight, "the machinery faltered," quipped *Newsweek*, "never Glenn."[103] Reporters offered a similar description of Cooper's flight. Cooper had to "take over," *Time* proclaimed, "when the best equipment that the best of science could provide failed."[104]

Ultimately, as columnist Raymond Moley pointed out in *Newsweek*, the space program was a triumph of "man," since "his ingenuity created the machines."[105] In late 1963, *Look* published a rare article on space, echoing Moley's sentiments. "As man makes more complex machines to do more unprecedented jobs," writer Ben Kocivar noted, "he must depend for ultimate success on man, himself."[106] Two writers at the *New York Times* went further, characterizing the debate over manned versus unmanned space exploration as a showdown between the individual and the machine. Two articles in particular stand out: an editorial immediately after Glenn's flight entitled "Let Man Take Over," and a piece in the *New York Times Magazine* by John Finney entitled "Astronauts Can't Be Automated." The articles reflect an underlying fear and frustration in the 1960s that machines were encroaching on the freedom and control of the individual.

One almost detects a sense of rage in some of Finney's reporting about the encroachment of the machine on the life of the individual. Ordinary taxpayers, explains Finney, accept the idea of placing instruments in space as a "logical extension of the now accepted practice of thrusting instruments into every conceivable place, explored or unexplored: down human throats, under penguin eggs, into the heart of the atom."[107] Proponents of unmanned space exploration use "the scientific dogma" that America's future "depends on basic research" to silence anyone "bold" enough to question the worth of sending expensive instruments into space. The successful flight of John Glenn changed all that. Days after the astronaut's flight, Finney maintained, Glenn stood on the same Turk Island that Christopher Columbus once explored and announced that in his triple orbit of the earth, America had succeeded in making man an indispensable part of the spacecraft. Finney labeled Glenn's remark "a turning point in history."[108] Not only did Glenn's flight stand as the first American orbital flight of the earth, it also stood as "a symbolic victory for man in the battle of man vs. machine," a battle now extending into "the infinite domain of space."[109]

The fiercest declaration of the sanctity of man came five days after Glenn's flight. The *New York Times* editorialist instructed readers that the lesson one should learn from Glenn's flight was that "we need not be ruled by machines."[110] Machines cannot think, the paper reminded its readers; they merely store thought. Human beings remain "masters of the inanimate world," the newspaper added, "and nothing we can make or imagine we can make will take dominion over us."[111] Human beings would not stand by helplessly as machines took over. On the contrary, the editorialist boldly announced, "we have control of our world and its future."[112] People should reject a belief in an "automatic stream of history" over which they have no control. The paper claimed just the opposite, again underscoring the value of individualism: "Let man arise as an individual, working with other individuals but not committed to the machines of blind mass reaction."[113]

Although the vast majority of journalists highlighted the struggle between humanity and machinery, a few attempted to reconcile the traditional American hero with the modern world. One finds the best example of this in descriptions of John Glenn. Writers took Glenn's "frontier" stoicism to the extreme, asserting that he functioned like a machine. *Time* stated that Glenn saw himself as "another piece of the machinery in the system."[114] *Newsweek* went so far as to entitle one of its articles "John Glenn: One Machine That Worked Without Flaw."[115] The *New York Times* characterized Glenn as blending with the machine: Glenn "epitomizes a giant step in that constant, driving process to blend the human being and the machine into a unit of high harmony."[116]

Describing an astronaut in traditional Puritan and pioneer terms, Americans could read into their history the necessity of looking back to their frontier past. Viewing an astronaut as a machine or as a partner with the machine, Americans could also view their history in terms of progress into the technological future. The views of the astronaut as a machine, as a partner with the machine, and as superior to the machine highlight America's conflict over the relationship between man and machine in the early 1960s. By celebrating the technological accomplishment of the machine, one lessens the impact of America's past as the source of its strength. By emphasizing the American past as the source of America's strength, one negates the progress of the increasingly mechanized future. A few writers of the time struggled to reconcile the two views, attempting to enable Americans to adopt both perspectives.

News coverage of the space program in the 1960s played a crucial role in shaping the perceptions of the American public. In examining the news coverage of American space exploration from 1959 to 1963, one finds that the vast majority of the stories, and the way journalists reported them, reflect three enduring news values: small-town pastoralism, altruistic democracy, and individualism. The first two of these values help account for much of the negative coverage of the space program: reports that the program had gotten too "big" or that selfish political concerns had taken precedence over "the public interest." But far more often, the media's attraction to stories of rugged individualism led to celebratory stories about the space program. The media's worship of the individual, moreover, created a bias in the controversy over a manned space program versus an unmanned one.

Also, one cannot overlook the natural tendency of American journalists to root for the home team while in the midst of a propaganda race with the Soviets. In part, this loyalty may explain their failure to scrutinize claims made by the administration, NASA, or the astronauts. Journalists, like other members of society, are not immune to the fears and uncertainty of the nation during times of national turmoil. The celebration of the space program reflected the media's desire to renew its faith in the American dream. As Raymond Moley of *Newsweek* observed, "We need this renewal of faith even more than we need to reach the moon."[117] At the same time, this desire for a renewal of faith led to a focus on the astronaut and his personal qualities—and on a mythology of the past—that may have distracted America's attention from the political, technical, and economic controversies surrounding the space program's future.

Besides journalists' concern with national pride, one cannot overlook the influence of the pocketbook. *Newsweek* understood the inherent human interest in manned space exploration. "No satellite, no matter how ingenious or scientifically valuable," the magazine observed, "can match the ageless human drama of the individual—solitary, questing, vulnerable—facing the unknown."[118] Simply stated, manned space exploration would sell more magazines and newspapers than unmanned exploration would have sold. William Boot, trying to explain the media's inability to see the warning signs of an impending disaster prior to the *Challenger* disaster, reminded the public that journalists have "a vested interest" in manned space flight. "Man-in-space," Boot declares, "makes for a much more readable—or viewable—story than machines."[119]

Concerns for pride and pocketbook, coupled with the media's reverence for individualism, suggest that the American media were predisposed to favor the administration's program to send a *man* to the moon. Such concerns may have impaired the media's ability to make a fair and accurate assessment of unmanned space flight and may have prompted the media instead to place an undue emphasis on manned flight. In the next chapter, we see the media's predisposition for manned space flight and its concern with its pocketbook reach its peak in *Life's* coverage of the space program.

4

Life: NASA's Mouthpiece in the Popular Media

In attempting to understand and evaluate how rhetoric shapes perceptions of social and political events, one cannot overlook the mass media's role in public affairs. As David M. Berg writes, the modern person's knowledge of world events "is largely attributable to the mass media of communication."[1] The media of mass communication serve as "gatekeepers," limiting our exposure to but a small piece of a complex world. In addition, journalists shape our interpretations and evaluations of those people and events to which we are exposed.[2] Acknowledging their vital role in our civic affairs, Lloyd F. Bitzer calls journalists the "new rhetors of the twentieth century."[3] Bitzer proposes that this new "journalistic discourse" deserves "most careful analysis and research."[4]

Henry Robinson Luce, the founder and owner of *Life, Fortune,* and *Time,* contributed significantly to the role that journalism plays in modern society. Luce epitomized the journalist as rhetor. Considered by some the "most influential private citizen" in America during his lifetime, he revolutionized modern journalism.[5] Most important, Luce saw his magazines as vehicles for promoting patriotism in America.[6] Luce frequently blended facts and opinion in his magazines without distinguishing between "news" and "editorials." Frequently the "facts" gave way to the demands of his conservative political ideology. "Luce's printed version of what he felt events should have been," David Halberstam argues, "often obscured what they in fact had been."[7] W. A. Swanberg went so far as to call Luce "the world's most powerful unacknowledged political propagandist."[8] A fierce anti-Communist, Luce believed journalists should assist the American government in winning the cold war. Indeed, as revealed in a 1953 speech to the School of Journalism at the University of Oregon, Luce believed that the cold war could not be won at all without a strongly pro-American journalistic ethic: "All the qualities of command of President Eisenhower and all the thoughtful brilliance of John Foster Dulles will not win the Cold War—without us journalists."[9]

On 23 November 1936, Luce introduced *Life,* America's first modern picture magazine. By the late 1950s, he was seeking to change the editorial formula and purpose of the magazine. In a twenty-seven-page document entitled "LIFE: A Prospectus for the Sixties," Luce declared that *Life* would fulfill the need for a great magazine with a national purpose. He summarized the magazine's purpose under two heads: (1) winning the cold war and (2) creating a better America.[10] Even within Luce's stable of unorthodox magazines, *Life* was unique. Like other popular magazines, *Life* covered newsworthy events. Yet when it did so, more often than not it adopted a "human interest" angle. The 2 March 1962, issue, for example, provided a human interest angle on John Glenn's flight. Although the cover pictured Glenn after his flight, the caption read, "The Glenn Story Nobody Saw: At Home with Annie and the Kids While John Orbited the Earth." Inside, the reader finds an article by Loudon Wainwright on the family's reaction to Glenn's flight. Pictures and text recreate the family's reaction to the takeoff, the flight, and the successful return. Besides a nine-page article on the history and current popularity of the English language, the rest of the articles covered the unusual, the trivial, and the insignificant.[11] The issue carried an article, for example, on a New York man who crashed parties and social functions to meet celebrities, politicians, and foreign dignitaries. The edition also featured the aquatic feats of Bing Crosby's two-year-old daughter, the eccentricity of millionaire Huntington Hartford II, and a short biography of gun maker Sam Colt. The magazine also attempted to appeal to fashion-conscious women with an eight-page spread on the dresses of two young European designers: Roberto Capucci and Yves St. Laurent. Finally, on the last page, *Life* ran a trick photo in which three surfboards appeared to have sprouted legs.[12]

Above all, *Life*'s uniqueness stemmed from its pictures. The magazine, Dora Jane Hamblin recalls, always looked for stories with "great visual impact."[13] In his history of Time, Incorporated, Robert T. Elson observes that in *Life,* "words always yielded priority to pictures."[14] Because Luce built the magazine around pictures, it revered photographers and placed them on a pedestal. Hamblin, a former reporter for the magazine, said that the photographer was viewed as a "God."[15]

Luce deliberately made the magazine larger than other periodicals—thirteen and one-half by twenty-one inches when laid out flat—to accommodate big pictorial layouts. Luce even used special paper that would keep pictures clear. The magazine's cover always consisted of a color photograph, usually of the face of a popular political or entertainment figure. Although the magazine did have an editorial page, photos dominated the stories. It was not unusual for *Life* to devote four or five pages of a six-page story to stunning color photographs with bold captions. Often, the magazine gave two entire pages to one photograph. And Luce's gamble paid off. As Estelle Jussim writes, *Life* brought the art of mass documentary photographic journalism to the "apogee of perfection."[16]

In the genre of photographically illustrated pictorial press, the picture magazines *Life*, *Look*, and *National Geographic* were unique. In these magazines, photographs often forced exposition to the background. Unlike newspapers or other magazines that used individual photographs or photographic sequences, picture magazines covered events with the photographic story, or narrative. *Life*'s editors understood, and exploited, the power of the photographic story. According to Maitland Edey, the magazine's editors learned early in the 1930s that they could arrange a collection of photos on a single theme "to convey a mood, deliver information, tell a story," in a way a single picture could not. Edey adds that picture selection and arrangement were all important in relating a story.[17]

The narrative form of the photographic essay may help to explain the success of picture magazines. One theory suggests that a key factor determining what news stories readers will choose is identification. Clearly, stories with pictures make the identification process easier for readers, especially photographs of people. Also, as Robert Rhode and Floyd McCall suggest, a reader gets the "strongest sense of identification—or sense of participation—when the story progresses from a beginning through a series of related developments to a definite ending."[18] In short, the reader identifies with the narrative form.

Halberstam proposes that *Life* was "less political, more open" than Luce's other publications. Its photos made it "tied to events themselves rather than to interpretation of events,"[19] he argues. But Henry Luce certainly did not see things that way. Although Luce knew little about photography, Halberstam has written, Luce sensed the "drama inherent in it" and believed readers wanted it.[20] Luce viewed the photograph as a significant instrument of patriotic journalism. Speaking to a group in Williamstown, Massachusetts, in

1937, Luce called the photograph "the most important instrument of journalism" since the printing press.[21] Through photographs, Luce could counteract what he saw as journalism's tendency to focus on the negative; he could use photos to present a positive, idealized portrait of America. In concluding his Williamstown speech, he labeled the photo "an extraordinary instrument for correcting that really inherent evil in journalism which is its unbalance between good news and the bad."[22]

Life's purpose and character thus seemed in sharp contrast to today's magazines. Unlike modern newsmagazines such as *Time* or *Newsweek*, *Life* made little effort at in-depth synthesis and analysis of events. Letting its pictures tell most of the story, it was primarily a visual medium—more like television news than *Time* or *Newsweek*. Neither did *Life* resemble today's magazines of opinion, such as *National Review* or *New Republic*. Since it carried editorials, *Life*'s articles presumably reported "news." Clearly, however, *Life* defined news much differently from most modern newsmagazines. In short, *Life* was an amalgam: a newsmagazine, an opinion magazine, an entertainment magazine, and television news all rolled into one.

In the midst of the emergence of adversarial journalism, *Life* retained its partnership with government. While in 1961 *New York Times* writer Homer Bigart produced articles critical of the Vietnam enterprise, *Life* ran articles supporting South Vietnam and American involvement.[23] The relationship between *Life* and the government went well beyond observing certain "niceties" and even beyond Luce's usual unbridled patriotism. *Life* actually had a contract with the government, specifically NASA, giving the magazine exclusive rights to the astronauts' "personal stories." In mid-1959, NASA invited the press to bid for the contract, and *Life* won. Other reporters cried foul, claiming that *Life* had squeezed them out with its checkbook.[24] NASA and *Life* held fast.

The unique, long-term relationship between *Life* and NASA deserves close scrutiny. How did this intimate relationship between one of the best-selling and most influential magazines in America and a government agency come about? And how did it affect coverage of the astronauts and the space program? What difference did it make in Americans' perceptions of the space program? Elaborating upon the Kennedy administration's frontier rhetoric, *Life* depicted America's manned space flight efforts in narrative form as a frontier adventure story. Furthermore, as I will show, *Life*'s coverage of the space program was persuasive and unique because it conveyed the narrative in both prose and photographs and because the interaction of the two visual media demanded readers' active participation as did no other medium covering the story.

Forging a Unique Relationship

Both NASA and *Life* had strong motives for establishing an exclusive contract for the coverage of Project Mercury. NASA's public affairs chief, Walter Bonney, came up with the idea of selling the astronauts' personal stories. Bonney wanted to negotiate a single contract for all of the astronauts. He feared that allowing them to make individual contracts would lead to competition among the astronauts and different publications for the most lucrative contract. In addition, a single contract with a single magazine would make it easier to control the coverage. Bonney contacted Washington lawyer C. Leo Deorsey, whose clients included Arthur Godfrey and Edward R. Murrow, to negotiate the deal. Deorsey agreed to represent the astronauts under the condition that he would not be paid for his services. NASA also decided to control the voices of the astronauts' wives. The agency would include them in the contract. According to Loudon Wainwright, Deorsey started bidding for the contract at $500,000. *Life* met the minimum bid and won the contract, which would last for the duration of the Mercury Project.[25] The astronauts' families, paid in installments after each flight, divided the money evenly among themselves.

The contract stipulated that the astronauts, "singly or collectively," could agree to sell their "personal stories, including rights in literary work, motion pictures, radio and television productions."[26] Although allowing the sale, NASA kept tight control of the astronauts' messages. Not only did the agency forbid the astronauts to appear on television, radio, and in motion pictures without prior approval of NASA's director of Public Information, but it also required the director's approval to "publish, or collaborate in the publication of, writing of any kind."[27] In practice, the provision allowed NASA to control any material written by the astronauts.

The personal stories of the astronauts seemed perfectly suited to Luce's *Life*. First, the inherent drama of the program and the possibility for stunning pictorials made the astronauts' stories perfect for Luce's brand of photojournalism. From its first issue on 23 November 1936, according to Dora Jane Hamblin, *Life* "had been fascinated by spectacle."[28] Second, the stories had tremendous potential as a weapon in fighting the cold war. Positive portrayals of the astronauts and the space program would help offset the tremendous propaganda successes the Soviets achieved with their space shots. No other story seemed better suited to Luce's conception of *Life*. Loudon Wainwright, who, along with military affairs editor John Dille, wrote most of the magazine's articles about Project Mercury, refers to the astronauts as "precisely the sort of journalistic commodity" that *Life* editors felt they "simply had to buy."[29] Finally, the magazine may have purchased the stories for a more pragmatic reason: to boost sagging sales. Although the magazine accounted for over half of corporate revenues in the 1950s, Robert T. Elson observes, by 1958 its numerical lead over its competitors had diminished and newsstand and advertising sales had slowed.[30] With television, radio, newspapers, and other large-circulation magazines closing in on the leader, *Life* was not only fighting the cold war, it was fighting for its journalistic life. Some journalists initially expressed concern about the contract. Although the magazine had routinely acquired exclusive pictures and accounts without controversy prior to the NASA contract, buying the astronauts' stories "set off all alarms."[31] For *Life*, the potential benefits of the contract apparently far outweighed any potential damage from negative publicity.

NASA likewise had good motives for entering the contract with *Life*. During the early manned space program in the late 1950s and early 1960s, *Life* was one of the best-selling and most influential sources of news in America. From 1958 to 1960, *Life* was the best-selling weekly in the country. Only the monthly magazine *Reader's Digest* had a larger circulation.[32] Although *Life*'s sales slowed from 1961 to 1963, it still remained one of the five best-selling magazines in America.[33] Circulation figures, however, do not tell the whole story. Market researchers estimated that each of the 6 million to 7 million copies of the magazine sold weekly between the years 1959 to 1963 were "read by five persons," not counting those who picked up the magazine in doctor's and dentist's offices or public library reading rooms.[34] Perhaps President John F. Kennedy's attempt to court *Life*'s favorable coverage stands as the greatest testament to the magazine's power. Kennedy, wrote David Halberstam, saw *Life* as a "key to the independent center." In the "pre-television era," Kennedy viewed Time-Life as "the most influential instrument in the country."[35]

NASA saw *Life* as a vehicle for popularizing the space program. Loudon Wainwright writes, "As always, NASA needed public awareness and acceptance to keep its programs going—and viewed favorably by Congress. *Life* was a superb vehicle for that."[36] The magazine's articles, notes Edwin Diamond, could give the Mercury astronauts a sort of "total and intimate publicity" that even "journalists of Lindbergh's day did not achieve."[37] The magazine's stunning pictorial layouts would dramatize NASA's rhetoric. *New York Times* writer John W. Finney points to another reason NASA found the contract attractive. The stories, carried in a commercial magazine, would carry "far greater propaganda impact" than "official reports" released by the U.S. Information Agency.[38] Unlike government publications, *Life* had the appearance of an objective weekly magazine.

The contract lasted the duration of Project Mercury, from 1959 to 1963. Not only did Time, Incorporated, secure exclusive rights to the astronauts' "personal stories" for publication in *Life*, it also gained all book rights. In 1961, it subcontracted with Golden Press to publish *The Astronauts: Pioneers in Space*. The book, aimed at young adults, included an introduction and conclusion by Loudon Wainwright and a chapter supposedly written by each astronaut. In 1962, Time, Incorporated, published *We Seven* through its own publishing company, Simon and Schuster. Written by the "astronauts themselves," the book consisted of an introduction by *Life* writer John Dille and numerous chapters written by the astronauts. But throughout the contract, the main focus remained on articles about the astronauts, their flights, and the space program in *Life* magazine.

Before any of the astronauts had actually gone on a mission, *Life* devoted approximately eighty pages to them and placed them or their wives on the magazine's cover three times.[39] From their first appearance on *Life*'s cover on 20 April 1959 to Gordon Cooper's appearance on 31 May 1963 the astronauts or their wives graced the magazine's front page twelve times. Moreover, *Life* devoted the cover of its 27 September 1963 to the nine new Apollo astronauts. The magazine's cover selection reinforced the primacy of human pilots. Prior to the first manned flight, for example, *Life* highlighted the flights of monkeys. In 1959, *Life* placed Able and Baker, two rhesus monkeys who had completed a space flight, on the cover and ran a lengthy article on them and their flight. In 1961, three months before Alan Shepard's flight, Ham, a chimpanzee who had completed a suborbital flight, graced the cover of the magazine. The featured article described Ham as a heroic "astrochimp." When another chimpanzee named Enos orbited the earth twice months before Glenn's flight, however, not only did *Life* not place the chimp on its cover, but it also failed to devote much attention to the feat. The magazine placed Enos's picture on page 2 with a small paragraph describing the flight. Besides displaying the astronauts on its cover, *Life* published over seventy articles on the astronauts and their wives from 1959 to 1963. The majority of articles came in groups immediately before and after the flights. Following the flight, the magazine usually printed its own story on the flight, the astronaut's reaction to the flight, and his wife's reaction.

NASA's public relations department took steps to ensure a positive image of both the astronauts and the manned space program. Edwin Diamond went so far as to call the original press conference remarks "the last natural expressions" NASA would permit the public to hear from the Mercury astronauts.[40] As part of its contract, NASA included a clause that specified that the astronauts and NASA had to approve all "personal stories" carrying astronaut bylines. Loudon Wainwright wrote many of the early astronaut-signed articles. Later the astronauts worked with *Life* writers in composing the articles, an arrangement that made many of the writers uncomfortable.[41] Wainwright explains the control NASA enjoyed. Both the astronauts and NASA's "cold-eyed hero shapers," Wainwright discloses, "had approval rights over most of the copy that *Life* churned out on this big running story."[42] If an astronaut did not like a statement he himself had made, the astronaut would have the ghosts change the comment even though he had not made such a statement. John Glenn's biographer, Frank Van Riper, proposes that the contract produced a "uniform picture" of the astronauts. "Anything that was not standard Apple Pie America," he adds, "did not see print."[43] The magazine's writers understood that they were playing a game of deception. Wainwright calls the magazine's coverage a "kind of cover-up" designed to inflate and smooth out the astronauts' image.[44] "Our principal job," he recalls, "would be the depiction—the enhancement really—of a winning astronaut image."[45] And the magazine would do so with both photographs and words.

Combining Photographs and Prose

Among the mass media, notes Kuo-jen Tsang, "news pictures have probably received the least attention."[46] To understand fully the power of *Life*'s coverage of the space program, one must understand the nature of photojournalism, both words and photographs. Wilson Hicks, executive editor of *Life* during the 1930s and 1940s, describes photojournalism's intent as creating a "oneness of communicative result" through the "dissimilar" visual mediums of words and photographs. Photographs and text combined, he adds, are much more effective than either one alone in conveying an idea or a story.[47] At the heart of photojournalism's persuasive power is the interaction of prose and photographs, which requires the active participation of its readers to make the combination meaningful. This interactive process is one of the things that distinguishes photojournalism from television. Another distinction may be in the nature of the images. Susan Sontag proposes that a photograph may be more "memorable" than moving images because it is a "neat slice of time, not a flow." Whereas a photograph honors a "privileged moment," each of television's images cancels its predecessor, each being part of an underselected stream.[48] Photographs have inherent strengths and weaknesses as a communicative medium. Studies demonstrate that a newspaper photograph increases reader interest.[49] One explanation for this may stem from the way people treat a photograph. Readers consider a photo as "incontrovertible proof" that a given event happened.[50] Photographs document events; they serve as evidence. As André Bazin observes, photography has a credibility absent from other picture making because of its "objective" nature. "In spite of any objections our critical spirit may offer," Bazin proposes, "we are forced to accept as real the existence of the object reproduced, actually represented, set before us, that is to say, in time and space."[51] Photographs fix images in one's mind that are hard to forget or refute. Yet photographs themselves are inherently weak in meaning. Paul Hightower finds only a modest agreement (31 percent) between photographer and viewers about the meaning of a photograph.[52] A photograph's strength—its ability to capture or freeze a particular moment in time—makes it weak in explanatory power.

A photograph's lack of explanatory power stems from two main causes. First, if a reader is to understand, he or she must have a context. Photos take events out of context. In freezing a moment, a photo sacrifices the context in which it is found. And without a context, a reader will find it extremely hard to understand how to interpret or evaluate a photograph. Second, as Neil Postman argues, since photographs speak about particularities (about not women but a woman), they are unable to offer an idea or a concept, unable to make an arguable proposition.[53] Thus accompanying text becomes crucial in making photographs comprehensible and meaningful.

Devoid of context and speaking in particularities, a photograph forces the reader's active participation in trying to decipher, to make meaningful, its content. Claiming that photographs themselves cannot explain anything, Sontag calls them "inexhaustible invitations to deduction, speculation, and fantasy."[54] The visual media of photographs and words function as question and answer. The photograph cries out for an interpretation, which the words invariably supply. Jean Berger and Jean Mohr explain the power of the interaction of the two visual media: "The photograph, irrefutable as evidence but weak in meaning, is given a meaning by the words. And the words, which by themselves remain at the level of generalization, are given specific authenticity by the irrefutability of the photograph. Together the two then become very powerful; an open question appears to have been fully answered."[55] *Life*'s coverage of Project Mercury offers a case in point.

Life Promotes NASA's Public Relations

With few exceptions, *Life*'s articles and editorials echoed the Kennedy administration's arguments for pursuing a manned space flight program. When the program came under serious attack in 1963, *Life*'s editor Edward Thompson wrote an editorial supporting the administration's efforts. Pondering whether the country should spend $20 billion to send a *man* to the moon and citing opponents' criticisms of the program, Thompson declared that America must continue to support Project Apollo. The program, he added, needs a *man* in space because of his ability to adapt to unexpected situations.[56] Late in 1962, John Dille excoriated Congress for making the Kennedy administration "beg in public for their appropriations."[57] Dille lashed out, blaming Congress for NASA's lack of success in manned space flight. The Soviets, Dille said, do not have to "dilute" their space effort by sending numerous unmanned "gadgets" to the moon "to pick up pebbles" and determine whether its "surface will hold a man."[58]

Manned space flight, the magazine proposed, would improve the country's prestige and stimulate America's economy. Late in 1962, for example, Paul Mandel wrote that America should get to the moon first to win "plain, old-fashioned national prestige."[59] While improving America's prestige, the moon shot would boost industrial output and stimulate the economy. In 1959, Thompson wrote that man "must get into space" to gain its "beneficent by-products."[60] In another editorial in 1962, Thompson pointed to various "by-products" in "power packaging and metallurgy" already generated by the space program. Perhaps in the most cautious note in any of *Life*'s coverage of the space program, Thompson proposed that although space exploration had unlimited industrial "potential," its growth would place "severe strains" on the economy. Space spending will benefit some parts of the economy while retarding others, he added. Although cautious, he proposed that America could pay for space exploration if it "slimmed-down" its economy, making it much more "competitive and efficient."[61]

Life stressed the most persuasive argument for sending a man to the moon: national defense. The Soviets had different "motives" for going into space, John Dille warned. They view space as more than a "scientific playground" or a place to gain "prestige." Wishing to exploit the military potential of space, the Soviets put their "best military brains" to work on their projects.[62] Thompson emphasized the moon's importance to America's future: "Who controls the moon may control the political lot of mankind."[63] Pointing to its potential use as a missile site, Paul Mandel underlined the "military reasons for hurrying to the moon."[64] Thompson went even further in an editorial in late 1962. He felt a sense of "urgency" in countering the Soviets' growing military role in space, proposing that the Kennedy administration should give the military a "stronger voice in the space program."[65] He seemed to sum up *Life*'s position on a manned moon shot well when he declared, "It matters little whether we call this a military or civilian project. It is an American necessity."[66]

Life provided numerous arguments for a manned, American moon shot. Most of its coverage focused on portraying the astronauts, their flights, and the manned moon mission in narrative form as a frontier adventure. From the beginning of its coverage of Project Mercury, *Life* called the astronauts' journey into space the "greatest adventure man has ever dared to take."[67] In *The Astronauts: Pioneers in Space*, Loudon Wainwright offered a similar assessment, referring to space exploration as an "adventure more exciting and more far-reaching than anything man has done before."[68] Similarly, John Dille characterized *We Seven* as "a personal narrative, full of suspense and adventure."[69] The astronauts, Dille added, made up the "raw material for the great adventure."[70]

More than an exciting adventure, the moon shot would fulfill man's destiny. In a 1961 editorial, Edward Thompson announced that man must go to the moon because his "destiny compels him to explore every unknown, every unattainable summit."[71] Writing about a manned lunar landing, Paul Mandel similarly proposed that if an astronaut did not reach the moon, Americans would not "fulfill man's destiny as an inquisitive being."[72] Just as the nation's forefathers were destined to venture into the West, Americans today were destined to venture into space. Thompson called America's commitment to space "a natural undertaking" for Americans, whom he labeled a "venturesome lot."[73] In late 1959, Thompson evoked the frontier mythology when he explained man's need to get into space by comparing it to the previous journeys of America's ancestors. "Man must get into space because it is there," he wrote, "just as America was there, and the U.S. West was there."[74]

The magazine often compared the moon shot to a foray into the Old West or to the voyage of Columbus. Attempting to elicit the powerful myth of the Old West, *Life* referred to the astronauts in its first article as pioneers. Introducing the astronauts' writings, John Dille characterized the men as "true pioneers" anxious "to mark the trail well" for others to follow.[75] The magazine also compared the astronauts' voyage to that of Columbus. Calling the astronauts the "first explorers" of space, Loudon Wainwright remarked that NASA's early space probings would be primitive, "like the voyages of Columbus and Magellan."[76] Edward Thompson attempted to deflect criticism of the space program in a 1963 editorial, comparing it to criticism of Columbus's journey. Any challenge produces "skeptics and naysayers," Thompson explained. Advisers tried to dissuade Isabella of Spain from financing Columbus's voyage, added Thompson, but the bold Isabella ventured into the "unknown" and changed history, turning her age into an era of unparalleled "exploration and discovery."[77]

Following the administration's lead, *Life* depicted the manned moon missions as a structurally coherent, frontier narrative. In the magazine's rhetoric, the missions possessed the four constituents of a frontier story: (1) an identifiable, conquerable geographic location that is (2) unknown and hostile and includes (3) a malevolent antagonist who is thwarted by (4) a heroic adventurer. Clearly, the moon presented a tangible objective.

In keeping with the story, *Life* portrayed the moon and outer space as hostile, unknown environments. John Dille warned that space, "with all of its unknowns—from plunging meteorites to searing radiation—was a hostile environment."[78] Even if the astronauts survive the hazardous voyage to the moon, Paul Mandel cautioned, "they will have to land unaided on a hostile surface."[79] Like frontier storytellers, *Life*'s writers added excitement to the moon shot by pointing to a human enemy lurking in the shadows. Thus the Soviets as human antagonists became crucial to the story's fidelity.

Doubtless, Americans viewed the Soviets as their enemy in space. In a 1962 article, Paul Mandel warned that NASA had "at least one tangible enemy—A Russian man-to-the-moon team."[80] Unlike the Americans, the Soviets would use space for militaristic ends. Edward Thompson warned that the Soviets viewed "control of space" as "control of earth."[81] *Life* staff writers criticized Soviet attempts to publicize Major Yuri Gagarin's flight. One writer attacked the Soviet account of Gagarin's flight as filled with "blatant and sometimes comic propaganda references."[82] Ironically, one could level the same criticism at some of *Life*'s coverage. The writer added that the Soviets had "carefully planned" the official celebration it showed the world.

The most important element of the frontier myth related to the nature of the heroic adventurer. The brave frontiersmen in space had to embody what Americans liked to believe were traditional American values, combining traits of both the Puritans and the pioneers. These rugged, fiercely independent men had to exert control over themselves and their environments. And like frontiersmen before them, the astronauts had to exemplify the Old West tension between individual and community.

With the exception of comments it made in one of its early articles, the magazine portrayed the astronauts as "ordinary supermen."[83] *Life*, observed Frank Van Riper, depicted the astronauts as "devoted family men, devout Christians, and above all, heroes."[84] Moreover, NASA may have been behind the positive image. Calling *Life* the "pacesetter," Van Riper characterized NASA itself as "the instigator" of the flattery.[85] The magazine worked diligently to create a preconceived image of the space flights and the astronauts.

From its first article on the astronauts, the magazine attempted to demonstrate the tension between individual and community. The magazine seemed unable to strike a balance between the two. *Life* declared that even though the astronauts came from the "same general mold," they remained "seven individuals." The article asserts that the astronauts must subordinate their "individual differences" to the larger interests of the group.[86] Loudon Wainwright hinted at the need for heroic individuals when he described the minutes before the first flight: "The closing hatch blots out the last of the outside light, and the astronaut is alone, prepared for history."[87] Like frontiersmen before them, the astronauts must succeed on their own against a hostile, unknown environment and the malevolent antagonist lurking within it.

Life emphasized the control that the astronauts would exert over their flights. John Dille characterized the astronauts' poise and actions as crucial elements to their success. Possessing "nerves of steel" and "devoid of emotional flaws," the astronauts had to remain "cool and resourceful under pressure."[88] In guiding their new machines through "hostile" environments, the astronauts would face unforeseen emergencies. At any moment, Dille added, their machines could fail, forcing them to become "masters of their own destiny."[89] Only seasoned test pilots, experienced in making calm, split-second decisions at high speeds and altitudes, could fit the role of astronaut. Dille called the astronauts "destined to fly" the new spacecrafts. Although destined for greatness, the astronauts must guard against excessive pride. They would have to be strong enough and "wise enough," Dille warned, to "sustain the pressures of public adulation when [they] returned home."[90]

The magazine followed the Kennedy administration's lead in covering the astronauts and their flights. *Life* writers seemed to adhere to the image the astronauts themselves wanted the world to see. Loudon Wainwright identified Alan Shepard, John Glenn, and Wally Schirra as the "most vigilant in the creation and protection of their images."[91] They wanted to appear confident, positive, and sensible. The astronauts, Wainwright recalled, instructed *Life* "ghosts" to portray the three as unemotional. The astronauts saw any perceived display of emotion as a sign of weakness, as a potential "loss of control and threat of breakdown."[92] Above all, stated Wainwright, the three astronauts wanted the world to view them as certain the whole enterprise would come out "just fine."

With few exceptions, the magazine portrayed the astronauts in glowing terms as heroic adventurers. In an article describing the three astronauts chosen for the first Mercury missions, Wainwright described Shepard as unemotional, detached.[93] Staff writers stressed Shepard's active control of his flight. Admitting that Shepard had flown "lower and slower" than Soviet Major Yuri Gagarin, *Life* echoed NASA's rhetoric, calling Gagarin a "passive passenger in an automatically controlled craft."[94] On the other hand, Shepard's more sophisticated craft gave him a "degree of protection and control not provided by Gagarin's." Shepard, the article continued, took over "personal control" of the capsule's pitch, yaw, and roll during the "peak" of the flight.[95]

The periodical did its best to offer a flattering picture of Gus Grissom. Wainwright's coverage of Grissom tried to deemphasize Grissom's shortcomings. The caption above Wainwright's article referred to the laconic Hoosier as "a quiet little fellow." Wainwright tried to account for Gus's lack of personal magnetism, calling him simply the "easiest" astronaut "to overlook."[96] John Dille also tried to offer a positive picture of Grissom. Early in the program, Grissom announced he was "'not the hero type.'" Three years later, Dille repeated and then reinterpreted Gus's comment, calling Grissom "completely indifferent to the spotlight" shone on the astronauts.[97] Dille even tried to paint Grissom's flight as a success. Admitting that the "freak accident" that had led to the sinking of Grissom's capsule caused Gus "embarrassment," Dille called the flight "so successful" that NASA needed "no further Redstone missions."[98] The Mercury astronauts, Dille added, believed that Grissom could not have been responsible for his capsule's loss. Dille's words, coming over a year after the incident, probably did little to restore Gus's image. Doubtless, it did little to hide the real reason why NASA canceled its one remaining Redstone flight: Soviet Cosmonaut Gherman Titov's seventeen orbits of the earth.

Glenn and his flight received the greatest amount of coverage. Glenn, Wainwright observed, attempted to create a positive image of the astronauts and the program. Glenn believed that the astronauts had "an enormous responsibility" to serve as "proper symbols of the nation's future."[99] Not surprisingly, *Life* portrayed Glenn in a positive light. Three weeks before his flight, the magazine explained why NASA did not select Glenn for its first flight: an anonymous high school friend of Glenn's suggested that the agency had been saving Glenn "'all along'" for its orbital flight.[100] *Life*'s description of Glenn's athletic prowess demonstrates its attempt to depict him as one of a special breed. Whenever Glenn's high school football team needed "sure yardage," they "invariably" ran through Glenn's "position at center."[101] During a speech delivered in his hometown after the flight, an embarrassed Glenn believed his audience would be "interested" in inaccurate media "accounts" of his "prowesses." The teachers, Glenn remarked, would be "very surprised" to learn that he had received "straight A's all through school," and his teammates would be amazed to find out that while he sat on the bench, the team ran "every play" through him whenever it "needed a few more yards."[102] Interestingly, only when addressing an audience that knew better did Glenn refute the media inaccuracies.

The magazine portrayed Glenn as a heroic adventurer, describing him as possessing the best qualities of the Puritans and the pioneers. Wainwright pointed to Glenn's religious devotion and self-control. Not only did he refer to Glenn as a "moral and strongly religious man," but Wainwright went further to conjure up images of the strict Puritans.[103] Wainwright characterized Glenn as a "stern and single-purposed" man who never swerved from his convictions.[104] Hinting at a masochistic tendency in the Puritans, Wainwright described Glenn's discipline and "self-denial" in getting ready for his mission. He described Glenn's approach to preparing for his flight as "decidedly ascetic." Proposing that Glenn believed one could "easily" endure pain for a "worthwhile" goal, Wainwright pointed out that to stay in shape Glenn "faithfully" ran two miles every working day. Like pilgrims before him, Glenn toiled long and hard in his "search for destiny."[105] Dora Jane Hamblin went further still, describing Glenn in messianic terms. Hamblin proposed that Glenn's stirring speech to Congress created "a deep silence, full of cleansing rejuvenating pride" in him, his family, and the nation.[106] Through Glenn, the nation, like a sinner, became cleansed and rejuvenated.

Life proposed that Glenn possessed the courage and patriotism of the early pioneers. Linking Glenn to earlier adventurers, Edward Thompson described Glenn's "personal courage" as "rare but constant" throughout history.[107] Like earlier Americans, Glenn had a great love for America. Hamblin wrote of Glenn's speech to Congress, "Its unabashed, star-spangled sincerity evoked the pride of nation of a far more innocent age."[108] Hamblin even tried to portray Glenn as a latter-day Davy Crockett, remarking that his comments "conjured up a vision not of stellar space but of the tough minded westward-moving wagon trains."[109] Glenn stood in a long line of rugged, courageous, and patriotic individualists who settled the country.

The magazine further emphasized Glenn's individualism when describing his isolation during his flight and the crucial role that Glenn himself played in making the flight successful. *Life* evoked the image of the heroic individual when it commented that during his flight, "no man was ever more alone."[110] Paul Mandel pointed to the necessity of having an individual pilot on the mission, stressing Glenn's active control over his spacecraft. Proposing that Glenn was a mere "passenger" during the first part of the flight, Mandel instructed that a systems failure in the craft forced Glenn to take over the "capsule's guidance" with a hand "control like those used in an airplane."[111] Mandel's comments, while praising Glenn, highlighted the shortcomings of machines. Glenn, Mandel wrote, "could change his mind and act where machinery performed slavishly and not well."[112] Mandel offered the strongest support for *manned* space flight when he stated that the voyage would have been "impossible" without Glenn's "intervention."

Although it gave Glenn widespread coverage, the periodical provided little coverage of Scott Carpenter, Wally Schirra, or their flights. Wainwright's description of Carpenter and his childhood presented one of the few deviations from the standard, apple-pie material written about the astronauts. Wainwright characterized Carpenter as "a hell raising kid."[113] Carpenter, he revealed, "stole" parts for his car and did "miserably in college." Unlike the other astronauts, Carpenter had nothing to lose by joining the Mercury program. Carpenter felt lucky NASA chose him as an astronaut, Wainwright observed, because when the agency selected him, his naval career "had reached a dead end."[114] Perhaps more surprising than Wainwright's remarks were Carpenter's own remarks. The astronaut admitted that for many years he was a " 'loser.' " Commenting on his high school days, Carpenter confessed to stealing from stores and being a general "no-good."[115]

Wainwright's article, portraying Carpenter as less than the all-American boy, caused a great stir in NASA. When NASA's "image protectors" read the article, recalled Wainwright, they became "upset."[116] Yet NASA could do little. Since the article carried Wainwright's byline and did not discuss Project Mercury, NASA's censors could not touch it. As stipulated in the contract, Carpenter had read and approved the piece. If the agency was mad at Wainwright, it was also angry with Carpenter for having approved such an article.[117] *Life*, Wainwright recalled, was "delighted" with Carpenter's candor, "probably because he sounded more like one of us than one of them."[118] It is possible that Wainwright and the other writers, upset about the strict censorship and perhaps feeling a bit guilty for going along with NASA's terms, may have jumped at the opportunity to write something humanizing about the astronauts.

The reporting of the final flights of Wally Schirra and Gordon Cooper returned to the positive characterizations of the earlier flights. John Dille depicted Schirra's mission as a textbook flight. The agency, Dille remarked, became confident in the success of its missions: "There was almost a total lack of butterflies in stomachs."[119] With increased criticism of manned flight in late 1962 and early 1963, *Life* emphasized the active control exerted by the astronaut. Cooper steered his flight "manually all the way," bringing his "bucking capsule" back to earth right on target.[120] Trying to elevate the astronauts, the magazine wrote that at the beginning of Mercury, "machinery stole the show." Cooper's flight, *Life* concluded, proved once again that "the real heroes" in space are "still men."[121] *Life*'s coverage of the astronauts, with few exceptions, portrayed them as all-American heroes. The magazine echoed the administration in portraying manned space exploration as a glorious frontier adventure. The periodical depicted the astronauts as frontier adventurers who had to overcome an enemy, the Soviets, while battling through an unknown, hostile environment to reach, and then to conquer, their ultimate objective: the moon. *Life*'s influence, moreover, did not stop with its own articles. It extended to the articles and books supposedly written by the astronauts themselves.

In the late 1950s and the early 1960s, journalism was not as self-critical as it is today. Few readers scrutinized or criticized *Life*'s coverage. Only when the magazine collaborated with the astronauts on books did it expose its coverage to criticism. *The Astronauts: Pioneers in Space*, a short book aimed at schoolchildren, received little attention from reviewers.[122] In contrast, the astronauts' chronicle of their space adventures, *We Seven*, did gain the attention of critics. Reviewers attacked the book's bias. Richard Witkin acknowledged the collaboration between *Life* and the astronauts, revealing that the magazine had given the astronauts "a great amount of help" in writing their accounts.[123] Reviewer Norbert Bernstein went further, writing that the astronauts' exciting prose made one suspect a "Luce ghost writer hovering nearby."[124] The reviewers attacked the "quasi-official account" of Project Mercury for its lack of critical appraisal. Witkin seemed gracious when he proposed that the astronauts "cast" the program in a more positive light than would "an objective historian."[125] R. C. Cowen came directly to the point, calling the book "weak" in "critical appraisal" of the country's manned space effort.[126]

The reviewers also noted the similarity between *We Seven* and the early magazine articles "written" by the astronauts. Bernstein warned that people who had read *Life*'s articles on space "may feel cheated" by the similarities. He added, "the book has many variations and additional text."[127] The material, supposedly written by the astronauts, did change from 1959 to 1962. From those changes, one can detect a conscious effort to emphasize the frontier motif and the active control exerted by the astronauts during their missions.

With increased emphasis on the frontier story and the need for a *manned* space flight program, the astronauts' articles, essentially controlled by NASA's public relations department but written by the magazine and the astronauts, changed over time to emphasize the astronauts' control. In the early articles of 1959 and 1960, and in their first book, the astronauts often called their missions "rides." They used various expressions, like the "Redstone ride" or "riding the capsule."[128] One caption atop a Gordon Cooper article in late 1960 read: "First Rocket We Will Ride." The tone of the later articles, and the second book, rarely refer to the missions as "rides." Instead, the astronauts refer to "flying."[129] Even the machine they would "fly" changed names. In the early works, the astronauts called their flying machines sometimes "rockets" but more usually "capsules."[130] One finds the beginning of the use of the term "spacecraft," which recalls navigable boats that one can pilot, in a 27 January 1961 article written by John Glenn.[131] Interestingly, in the 1962 book *We Seven*, many of the early articles of 1959 and 1960 are included, with some rewriting. In the new accounts, "capsule" is almost always replaced with the term "spacecraft."[132]

The subtle changes in language also appeared in the astronauts' depiction of how much control they exerted. In the early articles, the astronauts occasionally referred to themselves as "passengers." Gus Grissom, for example, once spoke of the "rocket and its passenger," while Gordon Cooper referred to the "astronaut passenger."[133] The astronauts drop this language in the 1962 retelling of their adventures. By 1962, the astronauts only used "passenger" in a negative sense.[134] In *We Seven*, the astronauts more frequently alluded to the control they exerted during their flights. Alan Shepard, Gus Grissom, John Glenn, and Scott Carpenter had all completed their flights by the time the book reached the public. Their accounts placed a much greater emphasis on the control that they exerted than did their earlier statements. Also more frontier terminology creeped into the 1962 retelling of their exploits. For the first time, Shepard and Glenn described space as a "hostile environment."[135] The astronauts used more direct analogies to the Old West and spoke more often of the unknown space environment and how "alone" they felt during their flights.[136] In short, as the administration pushed for more funds for its manned moon mission, the collaborative rhetoric between the astronauts and the magazine intensified. The retelling of the pilots' stories in *We Seven* placed greater emphasis on the frontier motif and on the nature and the role of the astronaut. Like the proverbial angler who, with each retelling of his or her story, embellishes the size of his or her catch, the writers of the astronauts' exploits embellished the necessity of putting man in space.

Besides covering the astronauts' glorious quest into the unknown reaches of space, *Life* also secured the personal stories of the astronauts' wives.[137] Attempting to appeal to its female readers by presenting the wives' reactions to their husbands' dangerous work, the magazine presented a pleasant picture of the wives. Edwin Diamond has written that in telling the wives' personal stories, *Life* "fixed" the wives' role as patient and waiting. If the "reality" did not "coincide with the image," Diamond observed, the magazine's writers would remake reality.[138]

Life lost little time exploiting the personal stories of the wives. The seven women appeared on the cover of the magazine's 21 September 1959 issue. Above the color picture of the smiling, smartly dressed wives appeared the caption: "Astronauts' Wives: Their Inner Thoughts, Worries." Each of the women—Anna Glenn, Rene Carpenter, Louise Shepard, Betty Grissom, Trudy Cooper, Jo Schirra, and Marjorie Slayton—"wrote" an article for the issue. Their stories supported the picture presented by NASA and the astronauts.

In the articles, the wives stressed their courage. When NASA first called John Glenn in for Project Mercury, Anna Glenn remembered, she was not "frightened."[139] Rene Carpenter offered a similar opinion: I had "no dark or foreboding feelings about having my husband prepare to rocket into space."[140] Like the rugged adventurers they married, the wives controlled their fear of the unknown. The wives' explanation of why they were not frightened, moreover, mirrored their husbands' reasons and painted a positive picture of NASA and the astronauts.

Like her husband, Anna Glenn gave up any doubt about the program after she learned more about it. Surely NASA would not undertake Project Mercury, Anna Glenn observed, "unless they knew what they were doing."[141] She had such faith in NASA that she could "laugh" at the thought that the agency would launch a rocket before it was perfectly safe. Trudy Cooper expressed a similar feeling: "I have an awful lot of faith in the engineering and technical skills of the people in this country."[142] The faith the wives expressed must have added to NASA's credibility. If the astronauts' wives trusted the agency, so should the general public.

The wives not only expressed faith in NASA, they also expressed faith in their husbands' abilities. Jo Schirra, for example, asserted that she and Wally considered test flying jets more "dangerous" than the "hazards" of space flight.[143] Similarly, Marjorie Slayton proposed that Donald's flying never made her "nervous" because he was such a professional that he could "handle" any "emergency."[144] The astronauts' expertise in flying, in controlling their crafts, presumably made the idea that they exerted control over their destinies far more credible.

Two of the wives evoked the spirit of long-ago adventures. Louise Shepard added a philosophical twist to her story when she suggested that man's willingness to take his first steps into space proved his readiness to "throw off more of the limitations he has put on himself."[145] Pointing out that many Americans wished to join in the space adventure, Shepard evoked the spirit of Columbus's voyage: "When Columbus set out to prove he wouldn't drop off the edge of the earth, there were others eager to go along with him."[146] Similarly, Trudy Cooper evoked the spirit of the old frontier when she wrote of Gordon's grandmother who went out West in 1895— "when pioneering took a lot of spirit."[147] When Gordon told his grandmother about the program, Trudy recalled, the grandmother got so excited that you would have thought "the Indian wars" had started again.

Four of the women wrote articles describing their feelings before, during, and after their husband's flights. Betty Grissom's article added further doubt to whether Gus had the "right stuff." She, too, wondered whether Gus had panicked. "I wondered if, accidentally, Gus might have done something wrong," she recalled. Betty went so far as to ask Gus, " 'It wasn't your fault, was it?' "[148] Although she announced that Gus said it was not his fault, the damage was done. If Grissom's own wife thought he had panicked, could anyone else doubt it? Unlike Betty Grissom's piece, Trudy Cooper's article reinforced Gordon's image, stressing the importance of his skill to his successful flight. When she first learned that Gordon had to "control his spacecraft himself" through reentry, she became scared. Yet she felt reassured when she learned from Deke Slayton that the "pilot, and not the automatic system, was the most important thing."[149] All he had to do now, she said, was "prove" his "ability as a pilot."[150]

The wives' personal stories, aimed at selling magazines to women, complemented their husbands' stories and demonstrated that they, too, came from good pioneer stock. Their statements about their husbands' flying abilities and NASA added greater credibility to the magazine's, and the administration's, portrayals of the astronauts and the program. Besides the wives' articles, the magazine's famous photography complemented the astronauts' story.

Making the Case Pictorially

By the early 1960s, *Life* had begun battling fiercely with television for advertising. Although television could broadcast live pictures of the special events upon which the magazine had previously made its living, television was still in its infancy. As Curtis Prendergast and

Geoffrey Colvin point out, "the network's evening news shows ran only 15 minutes."[151] Moreover, satellites had not yet begun beaming pictures from abroad. Still, competition from television may explain, in part, the periodical's eagerness to gain the personal stories of the astronauts, despite its tremendous cost and potential criticism from competitors.

From the start of Project Mercury, the magazine understood the program's appeal. In its second article on the astronauts, the periodical proclaimed that it would illustrate the story "in photographs as extraordinary to look at as the words are extraordinary to read."[152] *Life* assigned Ralph Morse to photograph the U.S. space program from its inception. He "became famous for his intricately engineered and technologically perfect pictures of launches, recoveries, and inflight training."[153] It was the interaction between text and photographs that made the magazine's depiction of Project Mercury as a frontier adventure so powerful.

NASA needed *Life*'s photographic coverage to make Project Mercury "real" and "believable" to the public. Prior to Project Mercury, the public viewed manned space travel mainly as science fiction. Photographs documented the events, allowing readers actually to "see" what for many may have been mysterious, unknown, or incomprehensible: astronauts, space ships, outer space, and space flight. The corresponding text allowed the magazine to explain the larger meaning of the particular astronaut and his mission as part of a frontier adventure, part of a human quest. Through the text, the reader came to understand that John Glenn represented not just the astronauts, not just Americans, but all people. His was the quest of humanity to free itself from its earthly shackles.

Life's editors arranged the photographs of the space program not only to coincide with the magazine's adventure story told in prose but also to create reader interest and identification and to document the events, thereby making the events and personalities "real" to its readers. The initial photograph of the seven astronauts separated them from everyone else, picturing them sitting behind a long desk at their first press conference.[154] The pictures accompanying the articles for the next two years showed the astronauts with their families and in various forms of training. Many of the early pictures showed the astronauts together with their wives and children. This reinforced their image as loving husbands and fathers. Morse also pictured the astronauts in training. The astronauts are shown, for example, in flight simulators, in a centrifuge machine that simulates the G forces the astronaut would feel during liftoff and reentry, and in the ocean with their space suits on, practicing recovery techniques.[155] In short, the pictures depict the astronauts preparing and training for their quest, their journey. Like the pioneers of old, the astronauts had to ready themselves for the battle.

The pictures accompanying the individual flights followed a pattern that illustrated and documented the prose describing their "adventures." The photographs begin by emphasizing the astronaut's isolation and separation from the larger community. The photographs show the astronaut eating his preflight meal, receiving his preflight physical, and putting on his space suit. Next, most of the photographic essays of the individual flights show the rocket sitting alone on the launch pad, followed by photographs of its actual, physical separation from earth; the photos record the rocket blasting off, flames belching from the boosters, white smoke everywhere. Often the periodical shows spectators looking on.

The pictorial narrative leaves the astronaut's encounter with his hostile environment and malevolent antagonists to the viewer's imagination. Instead, the magazine tries to convey the danger of the mission by providing drawings of fiery reentries and malfunctioning heat shields. In the later flights of John Glenn, Scott Carpenter, and Gordon Cooper, the astronauts' wives allowed Ralph Morse to photograph the family as it watched the flights on television at home. These pictures, particularly of the Glenn family, documented the danger of the flights. Morse, for example, shows the Glenn family's reaction at takeoff. Anna Glenn looks tense; her daughter, Lynn, cries, her hands covering her mouth.[156] On another page, the magazine placed three pictures side by side to enhance the drama. The first picture shows Glenn's white rocket, with flaming yellow boosters, against a brilliant blue sky. The middle picture shows Lynn with her eyes closed, her hands covering her face. Anna Glenn, mouth clenched, hands clasped together, stares at the television. In the final picture, the daughter wipes away her tears; Anna dons a big smile. The flight was over.[157] Morse uses this juxtaposition effectively in many of the pictorials to demonstrate the reaction of the families.

In the family's coverage, Morse captures well their agony and emphasizes the flight's danger and its drama. Moreover, the background for the photographs becomes important. In the private setting of home, one is not expected to put on a public face. Thus, one would expect a more genuine reaction from the family in their living room. Anyone skeptical of the danger of the flights need only look at the anguished faces of the family and the tears of joy and relief at the flight's end to erase any doubts.

Finally, the photographs depicted the astronaut's triumphant return and reintegration into the community. The photographs showed the space capsule dropping into the ocean. The astronaut had to survive one last trial in the dangerous, unknown waters of the

Atlantic. The editors highlighted the danger of the ocean landing with numerous shots of the capsules floating alone in the vast ocean. The magazine gave extended coverage to the sinking of Gus Grissom's capsule and his near-drowning. The astronaut, in short, would not be safe until his actual reintegration into society. Thus the recovery, and not merely a safe landing in the ocean, would signal the end of danger. In the recovery operation, photographs showed a helicopter either pulling the astronaut out of the water in a harness or bringing the entire capsule onto a waiting navy ship. Next, at least one photo showed the astronaut on the deck of a waiting ship surrounded by cheering sailors. Then the sequence showed the public response to the triumphant returning hero. The magazine often showed the astronaut opening fan mail, receiving medals from President Kennedy, riding in parades, or delivering speeches to a standing ovation in Congress. Invariably, the astronaut is shown with his wife and family, the trip completed.

In late 1963, *Life* ran an extended article on the new Apollo astronauts. As with their Mercury counterparts, *Life* shows the nine new astronauts training for their future journey into space. Reinforcing their all-American image, the periodical also shows the astronauts with their families.[158] Finally, the periodical shows the new astronauts together with the Mercury astronauts.[159] The torch had been passed to a new group of adventurers.

In depicting Project Mercury in words and photographs as a frontier adventure story, *Life* created a visual presumption in favor of a manned space program. As William Lewis has suggested, narratives are not refuted with "arguments." Rather, they must be supplanted by alternative narratives that ring truer for the audience.[160] The photographs, documenting and legitimizing manned space flight in concrete images that became fixed in the minds of Americans, created a tremendous burden for proponents of unmanned space flight. The familiar frontier narrative may have been so reassuring during the uncertainties of the cold war that no alternative narrative would have presented a serious challenge to it. Even if advocates of unmanned space exploration had been able to fashion a viable alternative narrative, however, they may still have failed because unmanned exploration did not have the visual power to challenge or supplant the familiar, identifiable images of the astronauts venturing into the final frontier of outer space. In short, images may, in some instances, play a critical role in determining which of the competing narratives audiences will choose. The more audiences identify with the images and the more the images document and support a particular narrative, the greater the chance that the images will influence an audience to accept the narrative.

With few exceptions, *Life* provided uncritical coverage of NASA's manned space program. Loudon Wainwright, looking back at his coverage of the program twenty-five years later, recalled his impressions at the beginning of Project Mercury. "What at first seemed like a great opportunity to work on one of the most interesting running stories of the age," Wainwright remembered, "began to seem a bit tainted with public relations."[161]

In 1962, *Life's* editor Edward Thompson wrote to NASA Administrator James Webb in an attempt to renew the magazine's contract with the astronauts, which ended with the completion of Project Mercury.[162] Webb's return letter attests to the influence the magazine's coverage had in helping the space program. Undoubtedly, Webb wrote, *Life* had done "an outstanding job in providing a basis for public understanding and acceptance of our national space program," particularly in manned space flight.[163] Both NASA and the magazine must have felt they benefited from the contract, since they renewed it the following year.[164]

Without doubt, *Life's* coverage of the manned moon mission, and the forum it provided the astronauts to tell their "personal stories," helped publicize the program, if not to influence readers' perceptions positively. The portrayal may have helped the administration gain public and congressional acceptance for Project Apollo. Still, an important question remains unanswered: why did the intimate relationship occur? There appear to be at least three reasons.

First, the nature of the press at the time permitted the relationship. Interestingly, the "vast majority of reporters did not think *Life's* coverage was wanting at the time."[165] They seemed upset, Frank Van Riper has written, only because *Life's* contract "deprived" them of the "chance to write the same type of article."[166] The press not only felt an obligation to observe certain niceties, it also felt an obligation to portray the astronauts "only in the most glowing terms."[167] Second, NASA, fighting for its bureaucratic life, wanted desperately to present a positive image to the public, thereby improving its chances for funding. In a 1 August 1962 memo, NASA's public affairs office enumerated the advantages and disadvantages of the contract. The office identified the reaction of the other news media as the greatest disadvantage.[168] The agency's control over its image seemed to outweigh any disadvantage. The contract, the office noted, "made it simple and easy for NASA to exercise its prerogative for reviewing all material before [being] published."[169] Finally, as Wainwright observed, in a "figurative sense," he and his fellow writers had "bought the story." There existed no "cold objectivity about national goals" or "the merits" of the project, he added. *Life* "had been charged with the protection of a valuable national asset."[170] And it labored long and hard to protect the astronauts and the program.

Life's coverage of the U.S. manned space program from 1959 to 1963, like the Kennedy administration's rhetoric, offered arguments for going to the moon and portrayed a manned lunar landing as a glorious frontier adventure. *Life* did not merely dress up NASA's public relations material with stunning pictures and bold captions. The magazine gave NASA a forum in which the astronauts could chronicle their adventures and sing the praises, and the necessity, of Project Apollo. The periodical also appealed to its women readers by presenting the personal stories of the astronauts' wives, who reinforced the administration's depiction of the astronauts and the program. Ralph Morse's spectacular photographs, moreover, complemented the magazine's text, presenting a pictorial narrative coinciding with the administration's adventure story.

Picture magazines, often combining photographic and prose narratives in their coverage, are unique in both the message they communicate and the reaction they evoke from readers. In forcing readers' active participation to make the stories meaningful, picture magazines differ from television. Moreover, in using photographic narratives, picture magazines differ from other photojournalistic coverage that relies on single photos or photographic sequences. *Life*'s coverage of Project Mercury was particularly powerful not only because it forced readers to participate actively but also because it elicited reader identification (1) through the content of the photographs (human beings in dangerous situations), (2) through the narrative form of the text, and (3) through the narrative form of the photographic essays.

In July 1970, *Life* finally quit its exclusive contract with the astronauts. The eleven-year relationship between a federal agency and one of the best-selling weekly magazines in America stands as a conspicuous example of the influence of the media on public affairs. Henry Luce epitomized the journalist as rhetor, and *Life*'s uncritical coverage of the space program undoubtedly had a strong influence on how Americans in the early 1960s viewed the program. The rhetoric of the Kennedy administration was similarly influential in the halls of Congress.

5

Congressional Space Committees: Overseers or Advocates?

"Congressional government," wrote Woodrow Wilson in 1884, "is Committee government."[1] Wilson's comment cuts to the heart of the legislative process. More genuine congressional deliberation occurs in the various congressional committees than on the floor of the Senate or the House. As John R. Fitzpatrick writes, Congress does "little, if any, debating" on the floor of the respective legislative bodies.[2] Rather, legislative proposals are most thoroughly investigated and debated within specialized committees, and the recommendations of those committees often determine the character and outcome of floor deliberations. Committee members typically go on record in support of or in opposition to legislative proposals well before the issues ever reach the stage of floor debate, and often enough other members of Congress publicly commit to positions before floor debates begin to determine the outcome in advance. Most floor debates are exercises, not in deliberation and decision making, but in justificatory rhetoric, or in public justification of decisions already reached. The truly deliberative rhetoric of the U.S. Congress occurs, not on the floor of the House or Senate, but within the vast committee structure of Congress.

For members of Congress, the frontier narrative advanced by the Kennedy administration and reinforced by the media not only became a way of understanding the space program and its surrounding events but also offered a specific program needing support—Project Apollo—that reaffirmed America's mythic identity during the uncertain years of the cold war.

Superficially, the committee structure of Congress seems to have engaged in thorough investigation of the Kennedy administration's space proposals—in analysis of both the general character and direction of the program and the budgetary requests. Four congressional committees oversaw the program: the House Committee on Science and Astronautics, the Senate Committee on Aeronautical and Space Sciences, and the appropriations committees of both bodies. The House Committee on Science and Astronautics stood as the most influential space committee. It conducted the most hearings, called the most witnesses, and heard the most testimony. The committee was divided into four subcommittees, with one holding immense power and influence over the direction of America's manned space program: the manned space flight subcommittee. It held more meetings, called more witnesses, and had responsibility for more of the space budget than any other space committee. In conducting authorization hearings for fiscal year 1964, for example, the manned space flight subcommittee alone held twenty-nine days of hearings, its proceedings filling 1,500 pages. In contrast, the entire investigation of the Senate's Committee on Aeronautical and Space Sciences that year lasted only eleven days and filled 1,100 pages. Overall, during authorization hearings and investigations from 1961 to 1963, the House and Senate space committees devoted approximately sixty days of hearings to manned space flight and heard testimony from nearly 200 witnesses. The combined hearings filled almost 5,000 pages. The Senate and the House appropriations committees also played a part in the budgetary process, but these committees were small, and their hearings brief. The two appropriations committees primarily studied the space committees' authorization figures and left investigations to the House and Senate space committees.

Yet a close examination of the rhetoric of these congressional hearings on space reveals that the congressional space committees never seriously questioned the need for *manned* space exploration, nor did they really question the massive economic commitment to space in general. NASA's budgets support this contention. The congressional space committees demonstrated their support for NASA's space programs by the size of the budgets they approved in 1961 and 1962: $1.7 billion and $3.7 billion, respectively. Not until 1963 did the committees finally make more than token cuts in the budget, trimming $600 million from NASA's requested budget. Even with the cuts, the space budget rose a whopping $1.4 billion from the previous year to $5.1 billion.

Still, one should not interpret the cuts as a sign of a truly critical posture by the committees. Numerous factors account for the cuts in 1963. First, the president's proposal for a joint moon mission caused not only confusion but also anger among many of the president's strongest congressional supporters. For years administration advocates on the space committees had sold space exploration as a matter of prestige and security in the competition with Russia. In one short speech to the United Nations, however, Kennedy undercut this appeal. The cuts may have reflected displeasure with Kennedy. Second, Congress approached an election year in 1963, and members of Congress became preoccupied with the issue of the economy. The space program, with a burgeoning budget and no natural constituency, became an easy target of the budget cutters. Finally, increased criticism of the moon shot by scientists and social liberals, numerous charges against NASA of mismanagement, waste, and patronage, and America's renewed sense of pride after its successful space flights and the showdown with the Soviets over Cuba, prompted leaders of congressional space committees to make some selective cuts to head off large, across-the-board cuts in NASA's programs.

The hearings of 1963 marked the first and last time Congress questioned the wisdom and efficacy of sending human beings into space. The examination, haphazard and halfhearted, never seriously threatened manned space exploration. Congress's cuts did not reflect a reevaluation of or displeasure with Project Apollo as much as they did a response to situational constraints tangentially related to a manned lunar landing.

No single event had greater influence on the view congressional committees took of the space program than Yuri Gagarin's orbital flight of the earth on 12 April 1961. The timing of the flight could not have been better for the administration. The flight took place just as the House Committee on Science and Astronautics began authorization hearings on NASA's budget for fiscal year 1962. The flight caused a virtual panic in Congress, especially within the House Committee on Science and Astronautics. A runaway mood swept over the space committee, and it did not abate until man set foot on the moon. The House committee demanded manned space spectaculars to match the Soviets', and it wanted them quickly. Thus when the committee might have been carefully examining the wisdom of manned space exploration, it was instead chastising NASA for not asking for more money, for not sharing its sense of urgency.

One can find evidence of the committee's mood in the comments and questions of its members. Representative James G. Fulton (R., Pa.) expressed his desire to speed up the program and beat the Soviets despite the testimony of NASA witnesses who said that the program could not move any faster. Although acknowledging the NASA scientists' testimony that they could not use any more money because they could not accelerate the program, Fulton pouted: "I want it faster." Fulton criticized NASA officials for not running their programs "around the clock." Instead of assuming the role of overseer, Fulton adopted the role of advocate. In questioning NASA's administrator, Fulton reminded Webb that he had said many times, "Tell us how much you need and we on this committee will authorize" it.[3] Fulton, far from scrutinizing NASA funds, attempted to force the agency to take more.

Other members of the committee also voiced their unqualified support. Some, like Fulton, chastised NASA for not doing enough. Representative Overton Brooks criticized NASA for dragging its feet. "Sometimes," Brooks maintained, "I question whether NASA has the proper sense of urgency of this program." Although he did not wish to scold NASA, he believed he voiced the committee's "feeling" on the issue.[4] Similarly, an impatient Victor Anfuso (D., N.Y.) demanded that NASA work faster. "I want to see what NASA says it is going to do in 10 years done in 5." Nothing short of a spectacular would do. Anfuso insisted on "some firsts" from NASA, "such as landing on the moon."[5]

Remarks by members of the committees suggest that they saw themselves more as advocates than overseers. Olin Teague, chairman of the influential manned space flight subcommittee, seemed particularly concerned with selling the program to Congress. On three occasions during the 1962 authorization hearings, Teague explained to his subcommittee members that it was their responsibility to "convince" the rest of the House that it should authorize the amount requested. "We have to justify this," he added, "on the basis of being laymen."[6] Teague continually looked for help in justifying NASA's budget to the House. In questioning two witnesses from industry sympathetic to the space program, for example, Teague asked them what they would say if they had the subcommittee's responsibility of presenting the bill on the floor of the House and persuading the other members of Congress to support the authorization.[7]

Still other members seemed to put their desire to justify the program ahead of their duty to scrutinize it. Chairman Brooks demonstrated his desire to help NASA justify its program when he instructed administrator Webb that the committee would get much "happier support financially" for NASA if the committee could show Americans the "civilian" benefits coming from the program.[8] Members also seemed concerned with justifying the program to their constituents back home. James Fulton, for example, asked a NASA official for help in justifying the Mariner spacecraft used for moon exploration. Admitting that he had at first thought he would support NASA, Fulton noted that he now had second thoughts because his constituents might question him about it. How would you "sell that gadget to the taxpayer?" inquired Fulton.[9] Likewise, Senators Allen J. Ellender (D., La.) and Spessard Holland (D., Fla.) pleaded with NASA officials to provide them with a justification of the moon shot that they could send to the numerous constituents who had written them letters asking them why the United States should embark on such a program.[10]

In this climate, not only did the committee fail to question the need for manned space exploration, it also failed to watch congressional purse strings. Committee members seemed ready to spend whatever was required to go to the moon. Representative David S. King (D., Utah), for example, claimed that he would favor any program, "regardless of the cost," that would place America in "the race to reach the Moon first."[11] The comment of Representative Jessica M. Weiss (R., N.Y.) to NASA administrator Webb testifies to the committee's runaway mood. It must be "very refreshing," Weiss observed, to come before a committee "anxious to give you more money than you seem to want." A delighted Webb responded, "We certainly appreciate this committee."[12]

Surprisingly, two committee members did try to maintain a critical posture. Representatives Perkins Bass (R., N.H.) and George Miller (D., Cal.) tried to slow the stampede. Bass, disagreeing with his fellow committee members, insisted that NASA had "shown a proper sense of urgency" in developing the space program. While others called for increased expenditures for NASA during the 1962 authorization hearings, Bass hoped that he would have the opportunity "to hear testimony and justification for the figures in the budget."[13] Likewise, George Miller reminded the committee that instead of engaging in histrionics about the Gagarin flight, the committee should get down to "brass tacks" and "hear some matters in support of the budget." Concerned about the size of NASA's budget, Miller declared, "I am not going to rubberstamp it in the interest of urgency."[14] These two voices, however, became lost in the clamor to speed up the program.[15] Gagarin's flight shocked and panicked committee members, prompting them to push for a manned assault on the moon even before they received justification for it. In short, the committee led the charge, exhorting the space agency to accelerate the program beyond NASA's stated capabilities. With such a fast and enthusiastic start, NASA's manned space programs gained such momentum that when criticism surfaced in late 1962 and early 1963, NASA overcame it without much loss of funding or support.

In the deliberations, the space committees focused on the political and defense concerns of the cold war and the key technical question whether man was necessary in space flight. In the minds of committee members, national defense stood as the strongest rationale for proceeding with a manned flight to the moon. Although some members were critical and raised doubts about the cold war and technical arguments, they were overwhelmed by committee members who made the positive case. The arguments the committee members advanced may be seen as chief concerns of those who understood the space program as a frontier narrative. The narrative highlights an antagonist capable of thwarting, or even killing, the pioneer. Clearly, the Soviets stood as the malevolent antagonists in America's adventure in space. Also, the frontier narrative places a premium on being first, on pioneering, on blazing a trail. The committee members' emphasis on the national pride that would be derived from being first and from beating the Soviets was consistent with the view of the space program as a frontier adventure, as was a concern for America's image abroad if it lost to the Soviets in its quest to conquer the moon.

Committee members justified the manned lunar landing as crucial for America's national security. In questions that committee members posed to witnesses and in remarks they made to one another, one discovers that national security carried a great deal of weight with committee members. Moreover, the military justification became more prevalent when, in 1963, critics began attacking Project Apollo. Although a few committee members questioned the military justification for sending a man to the *moon*, rarely did they question the military justification for *manned* space exploration, despite lukewarm support from Defense Department officials and scientists not related to NASA.

Gordon Allott (R., Colo.), the most vocal watchdog of the space program in the Senate, challenged the military value of placing a man on the moon. Questioning NASA's deputy administrator, Hugh Dryden, Allott asked whether in the foreseeable future "manned satellites" would have greater "military value" than landing three men on the moon. Senator Warren Magnuson (D., Wash.), chairman of the Appropriations Committee and an avid supporter of the program, answered for Dryden. "He says," Magnuson instructed, "you have to have the manned satellites before you can land on the Moon."[16] Allott's inquiry questioned the military value not of sending a man into space but only of sending him to the moon. The numerous questions that committee members asked about the military value of the program demonstrate their concern with the issue. Early in 1961, some committee members sought to use American national security to justify the moon shot. "Supposing we want to justify this program," Senator Leverett Saltonstall (R., Mass.) asked Dryden, can we do so by saying it will assist us in developing missiles "for the security of the country?"[17]

The influential House Subcommittee on Manned Space Flight, headed by Olin Teague, saw the national security issue as the program's greatest justification. Angered and confused by President Kennedy's call for a joint American-Soviet moon venture, Teague spoke for his subcommittee in a letter to Lawrence O'Brien, special assistant to President Kennedy. Ten of the eleven members on the subcommittee, Teague insisted, "support our manned space program almost completely on the basis of national defense and national security." Without the appeal to national security, he warned, "the subcommittee would have cut this budget in half."[18] The comments and questions of Teague, and those of other members of the subcommittee, testify to the centrality of the national security issue in their thinking about the space program.

Despite the testimony of scientists and Defense officials to the contrary, Teague maintained that Project Apollo served the interests of American national security. "I can't see how there can possibly be any doubt that there is a military mission in space," he asserted.[19] Other members, especially Republicans Richard L. Roudebush (Ind.), R. Walter Riehlman (N.Y.), and James Fulton (Pa.), supported Teague's hard-line stance on the defensive capabilities of the manned moon mission. Moreover, this shared, deep-rooted appeal, may explain the lack of partisanship in the committee. Teague, a Texas Democrat, and Roudebush, an Indiana Republican, for instance, shared a common distrust of the Soviets. "I don't have one iota of confidence" in the Soviets, Teague once declared in a 1963 authorization hearing. Roudebush seconded Teague's position. "I wouldn't trust the Russians any more than I could throw a dinosaur by the tail."[20]

The Republican congressmen showed their support in different ways. When critics attacked Project Apollo in 1963, for example, Representative Roudebush asked NASA administrator Webb questions that were actually statements about the national security issue. "Would you say it was a fair statement," Roudebush asked, "that we are not necessarily spending these millions and billions of dollars in space because we want to but because we have to?" Realizing the nature of the question, the astute Webb merely answered, "Yes, sir." Following up, Roudebush asked whether the administrator thought the American people "safer as a nation and individually due to our space efforts?" Again, Webb merely had to agree.[21]

Fulton and Riehlman, going even further, attacked Assistant Secretary of Defense John H. Rubel for not making bolder statements supporting the military benefits of manned space exploration. After Rubel declared that the Defense Department had not made a judgment on the necessity or nonnecessity of placing a man in space, Fulton chastised him for assuming that man does not have a military mission in space until the Defense Department has established one. Fulton maintained that he started from the opposite view, asking "'when does the man not become necessary?'" Aggravated by Rubel's testimony, Fulton added that he hoped General Bernard A. Schreiver, head of the Air Force Systems Command, had never said that man was "not necessary" in space exploration or that a "machine" could operate better without "judgment in it."[22] Likewise, Riehlman attacked Rubel for not making stronger statements about the military value of manned space flight, which Riehlman and his colleagues wished to use to justify the program. Riehlman felt perturbed by the Defense Department's failure to say to the "public" that it had "an interest in the space program." Observing that the committee would have to justify NASA's authorization bill before the House and the Appropriations Committee would scrutinize the bill, Riehlman complained that the DoD's "lukewarm" military support of the program made justification difficult. When, Riehlman asked, will committee members at least be able to say with "backing" from the Defense Department that a man in space program has some military value?[23] Riehlman's comments demonstrate his preoccupation not only with national security but also with justifying the program to Congress.

The Senate Committee on Aeronautical and Space Sciences also supported manned space exploration because of its relationship to national security. In hearing the testimony of the ten scientific witnesses before the committee, for instance, Senator John Stennis (D., Miss.) frequently questioned them about the program's military value. After criticizing Dr. Frederick Seitz's testimony for not making a strong enough statement about manned flight and national security, Stennis declared, "I very strongly believe there is a great military value to this space program." Later, Stennis justified the space program solely on the basis of its contributions to defense, calling passive inspection and surveillance satellites valuable enough to the military to justify "the cost of the entire program."[24] Interestingly, the satellites to which Stennis referred were made possible by unmanned rockets and had little to do with manned flight. One can see the concern for defense in the remarks of other committee members as well. Noticing that Dr. Martin Schwarzchild did not comment on either the space program's cost or its contribution to defense, Senator Stuart Symington (D., Mo.) proposed that his experience in Congress taught him that the only thing "more persuasive than the pocketbook" is "fear." Schwarzchild's testimony, Symington noted, addressed neither the program's cost nor its importance to defense.[25]

Not all members of the space committees, however, saw the manned lunar landing as indispensable to national security. Representative George Miller, chairman of the House Committee on Science and Astronautics, proposed that at present, he did not think anyone could say for certain that there is "a military need of a man in space."[26] Another committee chairman joined Miller in questioning the military value of sending a man to the moon. While members of his committee touted Project Apollo's contribution to the national security, Chairman Clinton Anderson (D., N.M.) disagreed, telling his fellow senators, "I am not so sure that there is tremendous military advantage."[27] Although the chairmen questioned the military value, they did not push the issue. Moreover, one rarely finds comments challenging the program's contribution to national security. From the comments and questions of committee members, it appears they especially fell back on the military argument when the program came under fire in 1963. Committee members seemed so concerned with justifying NASA's authorization bills that they pressed Defense Department officials to make comments that would help the committee sell the space program. The common concern for national security, moreover, may explain the lack of partisanship in the committees. Although scientists, military leaders, and committee members expressed reservations about the military value of a manned moon mission, the committees relied on it heavily to justify the program. Perhaps Representative Edward P. Boland (D., Mass.) best explained the committees' position with regard to national security when he remarked, "It is easy to sell defense in space."[28]

In the House committee's knee-jerk response to Gagarin's flight, it never stopped to question whether pride and prestige, two of the main reasons it cited for going to the moon, would be worth the expenditure. On only a few occasions did committee members ever question the prestige motives. The overwhelming majority of space committee members cited pride (or beating the Soviets in the race to the moon) and prestige (improving America's image at home and abroad) as major reasons for a manned lunar landing. Representative King, for example, viewed the moon shot as a race between the superpowers, a race he wanted to win. America hopes to get to the moon, he submitted, and "get there first" because the loser of the race will get "no prizes."[29] Actually, the members seemed less concerned with getting to the moon than they did with losing to the Soviets. Representative Fulton expressed this sentiment when he complained, "I am darned tired being second."[30] Chairman Overton Brooks agreed with Fulton. "My objective" in the space effort, Brooks revealed, "is to beat the Russians."[31]

Committee members also viewed the manned moon shot as a means of improving the image of the nation, which they believed the Soviets had tarnished with their manned space flights. During the debates over NASA appropriations in 1963, Senator Warren Magnuson reminded his fellow senators of the "element of prestige involved" in the program. "Let us not kid ourselves," he added, "this is part of it."[32] Representative J. Edgar Chenoweth (R., Colo.) believed the moon shot would help America regain some of its lost influence. By moving the space program ahead as "rapidly as possible," Chenoweth insisted, America can "recoup some of that prestige, some of that influence."[33] No one seemed more concerned with America's image than Representative Fulton. Speaking before the committee the day after Gagarin's flight, Fulton asked committee members whether they realized that they had the "responsibility of the way the capitalistic system looks to the world." His paranoia with maintaining America's image went to comical extremes. During authorization hearings in July 1961, Fulton proposed that since the moon stood as such a visible "propaganda symbol," the Russians might possibly explode a rocket filled with red dust on the Moon's surface, thereby turning the Moon red. Perhaps America should have a "blue project," he reasoned, to scatter "blue dust" on its surface, thereby making the moon red, white, and blue.[34]

The space committees saw manned space exploration as a means to an end: defending the nation, soothing their hurt pride, and improving America's image. To the extent that the members thought manned space flights supported these ends, the members seemed to support the program. They never actually questioned the defense or prestige value of manned flight or whether other activities or events, space related or not, would add even greater prestige and security to the country. Also, they never questioned whether American prestige needed such a boost. In the minds of the committee members, Gagarin's flight justified the expenditure.

In the hearings of congressional space committees, one occasionally finds members making remarks about the value of placing a man in space. Many enthusiastic supporters of the manned space program used the opportunities to question witnesses, particularly John Glenn, to offer their support of manned flight. Chairman Robert Kerr, for example, told Glenn that from his statement Kerr assumed America would make "greater progress," more quickly and more efficiently, with a "manned spacecraft" than with a spacecraft "controlled mechanically from the ground." Glenn responded that he was "sorry" he had not "put it in those words" himself.[35] Representative Roudebush set up Glenn by first insisting that Glenn's flight demonstrated man's ability to think and perform tasks while in space before asking Glenn if he felt it possible simply to "fly our future capsules and thus allow the pilot more capability?" Roudebush then allowed Glenn to make a long statement about the indispensability of man in space without once challenging anything Glenn said.[36] On a few occasions, committee members did challenge administration witnesses. The pointed questions usually revealed that the witnesses were misleading the committee. Senator Spessard Holland (D., Fla.), for example, challenged Glenn's characterization of the control he exerted during his flight. Holland noted that many Americans worked under the "misapprehension" that Glenn's manual controls enabled him to go where he wanted in space, whereas Holland understood that Glenn could not control his direction or speed, only the attitude or posture of the capsule. Glenn agreed with Holland, adding that he looked forward to the time when man could actually control the capsule and assume his "rightful place in space."[37] Likewise, in authorization hearings in 1961, Senator Warren Magnuson asked Hugh Dryden about the practical value of placing men on the moon. When Dryden answered that setting the goal required the nation to push itself in science and technology, Senator Gordon Allott followed up, asking, "But it has no practical use, that you can see, in itself?" Flustered, Dryden admitted it did not make any sense to him.[38] One rarely finds committee members taking this critical posture. Usually they accepted without question the need for manned flight. Part of the explanation may come from the excitement of the flights.

When committee members questioned witnesses, especially the astronauts, they usually got caught up in the excitement of the flights. Consequently, they often failed to scrutinize witnesses' testimony. Senator John Stennis seemed too busy praising John Glenn after his flight to ask any serious questions about the space program or the need for manned exploration. Asserting that Glenn personified "the very best in Americanism," Stennis added that Glenn's performance in space would make it much easier to appropriate money for NASA.[39] Occasionally, members seemed to get caught up in the excitement of the adventure story rhetoric. Dr. Colin Pittendrigh's testimony on the inherent adventure of space exploration, for example, impressed Chairman Anderson. Anderson seemed "particularly struck" with the professor's remark that society could not "perceive so great a challenge and not go after it." Catching himself, the chairman added that "of course," the committee was also "naturally interested" in what Pittendrigh said about the scientific aspects of the program.[40] Once a congressman used the frontier motif to comical effect. James Fulton evoked the frontier mythology in arguing against the building of a proposed auditorium at a NASA facility. "When the pioneers went out West," Fulton observed, "they didn't take auditoriums along to decide what to do next. I look at this," he added, "as a pioneering venture."[41]

In sum, committee members rarely challenged administration officials' claims that manned space exploration offered great advantages over instrumented flights in the early years of American space exploration. When they did challenge administration officials and representatives, the officials usually had to back away from their claims or clarify their positions. Although the congressional space committees rarely challenged the need for manned flight, they did, at times, criticize the country's space program.

In 1961 and 1962, NASA led a charmed life. Its budget grew from $1.7 billion to $3.7 billion in one year. Even though the space program had widespread support in Congress, it did have its detractors. A few committee members raised questions, particularly in 1963, about various aspects of the program. A few began questioning the scientific value of the program, while others attacked NASA's spending on public relations and the unequal distribution of space contracts. Members wanted to make sure that their home states received their fair share of the spoils. Overall, these criticisms were isolated cases, many times waged by the same members. Even the program's tremendous cost met with little opposition. One finds isolated instances of dissent. Representative Ben F. Jensen (R., Iowa), for example, felt a "little dubious," fearing that the program would put America in "bad financial straits."[42] Representative Joe L. Evins (D., Mass.) offered a similar concern. Although he felt "enthusiasm" for the program when NASA put an astronaut into orbit, Evins had a "different reaction" when he looked at the cost a "second time."[43] The criticisms, however, did not address the question of manned space exploration. In short, congressional space committees offered little serious challenge not only of the manned programs but also of the space program in general. A few committee members raised questions, but their criticisms were neither sustained over a period of time nor shared by a sizable number of their colleagues. The members seemed more concerned with discovering arguments with which they might justify the program to Congress and to the public.

Self-Interested Witnesses and Predictable Testimony

The space committees' failure to question the need for a man in American space exploration can be explained, in part, by the kinds of witnesses they invited to testify at committee meetings and hearings. Of the approximately 200 "experts" called before the space committees, nearly 140 were from NASA, 20 from the military, 15 from industry, 10 from universities, 6 from the Atomic Energy Commission, and 4 from the General Accounting Office. Virtually all had formal or informal ties to NASA. The government officials had a vested interest in the program, since many of their programs would directly benefit from increased spending on the space program. The industry experts also had a vested interest in the continued well-being of the program. The space committees, for example, called on representatives from companies like North American Aviation Corporation (prime contractor for the Apollo spacecraft), General Electric, Bell Telephone, and Thiokol Chemical Company. Each of these companies had huge contracts with NASA. As one observer of the legislative processes remarked, "Congress gets advice that is almost unanimously of a self-interested kind."[44] Thus it is not surprising that House members never challenged the administration's emphasis on a manned moon mission. Similarly, the Senate Committee on Aeronautical and Space Sciences called before them many of the same witnesses. When, in 1963, serious criticism against the space program began surfacing, Senator Margaret Chase Smith (R., Maine) suggested to the chairman of the Senate space committee that it hear independent, unbiased, technical opinion about the program. Consequently, the committee asked ten government, industrial, and academic scientists to testify on America's space goals. But only a few of these witnesses proved willing to challenge the prevailing consensus, perhaps because the majority also had a stake in promoting the administration's policies.

On 10 and 11 June 1963, the Senate Committee on Aeronautical and Space Sciences heard the testimony of an impressive group of scientists. Dr. Lee A. DuBridge, president of the California Institute of Technology, spoke to the scientific aspects of the program when he asserted that "competent scientists" could agree on what a scientific research program of space should entail. Disagreements, he added, come on the "relative validity and value" of nonscientific aspects.[45] Ironically, however, the scientists disagreed about the scientific worth of the program. Dr. Lloyd V. Berkner, president of the Graduate Research Center of the Southwest, and the chairman of the Space Sciences Board of the National Academy of Sciences, believed the space program offered great possibilities for scientific research. Berkner called the scientific objectives in space "real and powerful by themselves."[46] Dr. Colin S. Pittendrigh, professor of biology at Princeton University, also offered scientific justification for the space program. Pittendrigh insisted that the current space program could make great strides in discovering extraterrestrial life.[47]

Three of the ten witnesses, however, questioned the scientific value of the manned moon mission. Nobel Prize winner Dr. Harold Urey, professor of chemistry at the University of California, San Diego, assumed that Congress, in appropriating funds for the program, understood that it was "not justified in terms of the science that will be done."[48] One finds a similar opinion in the testimony of another Nobel Prize winner, Dr. Polykarp Kush, chairman of the Department of Physics at Columbia University. Calling the scientific objectives of the program "rather limited," Kush doubted whether the "purely scientific results" would be "reasonably commensurate with the investment."[49] Easily the most vocal critic of a manned moon mission, Dr. Philip Abelson, director of the geophysical lab at the Carnegie Institution and editor of *Science*, distinguished between the scientific value of unmanned and manned space exploration, proposing that the former had great scientific value. Abelson, however, attacked the manned missions. "Manned space exploration," he submitted, "has limited scientific value and has been accorded importance which is quite unrealistic."[50] Furthermore, he believed his opinion reflected that of the majority of scientists. In a straw poll of scientists he conducted for *Science*—scientists not "connected by self-interest" to NASA—Abelson found that only 3 of the 113 questioned favored the current manned space program.[51] Abelson refuted arguments that man would add to the missions scientifically, especially the notion that man could better deal with the unexpected. Abelson characterized man as a "poor scientific instrument" in space unless reinforced by instrumentation because his vision is limited. To make observations, man must rely on instruments already in his capsule. Therefore, Abelson concluded, "it would be necessary ahead of time to decide what feature of the unexpected should be anticipated."[52]

Although the witnesses disagreed on the scientific value of the missions, many of the scientists still supported the program. They typically left the realm of technical argument, shifting the ground of the debate to arguments more characteristic of the administration's rhetoric.[53] Dr. Berkner tried to offer justification for their shift. "Quite properly," Berkner noted, "scientists do not feel at ease in discussing the political, social, and economic implications of a program." The difficulty comes, he adds, when the "objectives of science" are used to make value judgments about America's national objectives. Berkner, however, insisted that he had to comment on these issues because "the national 'stakes' in space are of such dominating importance."[54]

Many of the scientists thus justified the program in terms of national prestige. Dr. Simon Ramo, vice chairman of the Board of Directors at Thompson Ramo Wooldridge, Inc., argued, "Let us not underestimate or be ashamed of an interest in science that is partly for prestige purposes."[55] He further identified prestige as part of the justification for *manned* space exploration. "A space program without man," Ramo observed, "has much less useful prestige appeal."[56] Dr. DuBridge offered similar testimony, proposing that a successful space program would certainly "elevate our prestige in the eyes of others," perhaps even making us "think more of ourselves."[57] Perhaps the most striking testimony came from Professor Urey. Stating that everyone wants to win a "race of some kind," he admitted he had an "interest" in seeing America "get to the moon first."[58]

The witnesses, although trained as chemists, biologists, physicists, and astronomers, also felt qualified to comment on the manned moon mission's contribution to national security. In all fairness, the committee solicited their comments. While a few scientists admitted that they lacked expertise to make informed judgments about military matters, most of the scientists offered their opinion. Again, however, the scientists disagreed. Dr. Frederick Seitz, president of the National Academy of Sciences, argued that the manned moon mission would enhance American national security. Seitz testified that the space program had "relevance to our military strength, both directly and indirectly."[59] Others saw things differently. Dr. Urey's testimony typifies the statements made by those who disagreed. The manned lunar program, Urey declared, "has no contribution to make to the national defense at all."[60] Dr. Abelson went further, stating that Project Apollo was even "detracting from our national security."[61] Still another group, while acknowledging its inability to find an application of manned flight to national security at present, maintained that the potential future risk that the Soviets might develop military applications necessitated a manned lunar program.[62]

The scientists testifying before Congress made their most striking departure from technical reasoning when justifying the manned space program on philosophical grounds.[63] "We must look upon this period of exploration of space," mused Dr. Frederick Seitz, "as an important part of the human journey applied to whatever meaning our own existence has."[64] The comment seemed more appropriate for a professor of philosophy than for the president of the National Academy of Sciences. In similar comments, Professor Colin Pittendrigh acknowledged the unconventional nature of his remarks. Scientific experiments in space, the Princeton biologist argued, offered the chance of obtaining a new perspective on "man's place in nature." Pittendrigh conceded that his remarks had "a far greater philosophic note than had been fashionable and, indeed, respectable in science for some time." But, he submitted, "I am quite deliberate in my emphasis on the philosophic note."[65]

Surprisingly, all but one of the scientists attempted to justify the program in terms consonant with the adventure-story rhetoric of the administration and much of the media. The witnesses frequently compared space exploration with the exploits of Columbus, Magellan, and the Wright brothers, and some cited the call to adventure as the program's greatest justification. Dr. Kush, a critic of the manned space program, nevertheless supported the program overall for the challenge it provided. "Man," Kush declared, "had no option but to climb mountains, sail over unknown and hazardous seas, explore the poles of the earth and learn to fly."[66] Remarkably, Kush concludes that this "urge" provides the "greatest justification" for space exploration. One might merely dismiss Kush's testimony as an isolated instance. But other witnesses also referred to this "innate characteristic" of man. Dr. Pittendrigh compared the space challenge to those of years gone by. Man must seize the challenge of space exploration, he proposed, just as people in the Elizabethan era responded to "similar adventures." Pittendrigh called the space program "justified" because it gave Americans a "sense of shared adventure and achievement."[67] Some scientists compared space exploration with other adventures in America's past. Princeton astronomer Dr. Martin Schwarzchild, for example, deemed the space age "entirely natural and quite in character with the other great pioneering phases of our past."[68]

The ten scientists who volunteered to testify on America's space goals before the Senate Committee on Aeronautical and Space Sciences disagreed on the scientific value of a manned lunar mission, but many still justified the program on nontechnical grounds. How does one account for their testimony? Part of the explanation relates to the composition of the group. Presumably, the committee desired the testimony of independent, unbiased scientists. These scientists, however, had strong ties to NASA. The day before the hearings, John W. Finney wrote in the *New York Times* that all but one of the scientists had "a direct financial interest in the space program," receiving either contracts or research grants from NASA or serving as "directors of institutions holding large contracts with the space agency."[69] Committee chairman Clinton Anderson (D., N.M.) cited the article during the hearing and questioned some of the scientists about their ties to the space effort. Dr. Pittendrigh began his testimony by stating that he had "no vested interest in the space program." When challenged, he admitted he received funds from NASA for part of his work.[70]

Two of the scientists seemed to have extremely close ties to NASA. Only a few weeks prior to testifying before the Senate's space committee, Dr. Frederick Seitz, president of the National Academy of Science, participated in NASA's third annual Conference on the Peaceful Uses of Outer Space, held in Chicago. His participation raises doubts about his objectivity. Perhaps Dr. Lloyd Berkner, a fellow member of the academy, had the strongest ties to NASA. A close friend of NASA administrator James Webb, Berkner chaired the Space Sciences Board of the National Academy of Sciences. In 1961, he successfully pressed the board to support Project Apollo. Thus, with one respected scientific group supporting manned space flight, the administration could dismiss scientific criticism of Project Apollo as a "legitimate disagreement" among scientists. The administration could thereby neutralize scientific critics and shift its case to purely political justifications for going to the moon.[71] One finds still further evidence of Berkner's ties to NASA. In 1961, he participated in NASA's first Conference on the Peaceful Uses of Outer Space. Thus, self-interest may have caused scientists to make claims and to advance arguments they normally would not have made. As Amitai Etzioni commented, "A desire to justify the moon race led some natural scientists, friends of NASA, to stray in their testimonies into the depths of the social sciences."[72]

Etzioni also speaks to a larger issue. Who judges the experts? In an increasingly technical world, members of Congress must rely more and more on the judgments of experts. As Etzioni points out, "when scientists testify as scientists, Congress expects them to act not as an interest group but as disinterested experts, evaluating issues on their scientific merits." Members of Congress tend to forget that scientists have personal interests and political views. Part of the problem stems from the ethos of scientists in American culture. One does not expect scientists to have political axes to grind, and if they do, rarely are those political views well known in Washington. Scientists rarely confess to their "own pecuniary" interests, or that of their subgroup. Their potential gains, like sitting on a prestigious research board or gaining increased federal funding for their branch of science, are hard to identify. Thus with scientists, members of Congress "let their political guard down."[73]

The problems the Senate space committee faced when seeking scientific testimony on America's space goals also indicate a larger problem of congressional decision making in an increasingly technological world. The testimony of the ten scientists before the Senate Committee on Aeronautical and Space Sciences in 1963 reached well beyond the technical arguments associated with science into the realm of public, political argument. Although this testimony may represent one striking, isolated instance, it does suggest a deeper problem: that scientists, like all people, have personal interests and political leanings.

Developing the Case for Project Apollo

The types of witnesses from whom the space committees heard partly explain why the committees never seriously challenged the need for a manned lunar program. The leadership of the space committees provides still another clue. From its inception in 1959, the House Committee on Science and Astronautics warmly and enthusiastically supported the national space program.[74] One can say the same for the other space committees. Part of the explanation for this reception comes from the leaders of the various committees. Not surprisingly, those with strong ties to the administration assumed leadership positions. The multibillion-dollar space program, moreover, gave the administration numerous opportunities to reward faithful service.

Representative Olin Teague (D., Tex.) chaired the manned space flight subcommittee, "the most glamorous, most senior, and most active subcommittee with the biggest budget and the greatest focus for publicity."[75] More important, the subcommittee made decisions about manned space exploration. Teague worked tirelessly to sell manned space flight to members of the House. As Ken Hechler (D., W.V.), who served on the House Committee on Science and Astronautics for eighteen years, has written, "Selling the space program to Congress was no easy task, and Teague and his subcommittee shouldered the heaviest share of the burden."[76] Moreover, Representative Albert Thomas (D., Tex.) chaired the House Subcommittee on Independent Offices, the body responsible for appropriating NASA's funds. It was no coincidence that both Teague and Thomas hailed from Texas, the state in which NASA chose to build its multimillion-dollar Manned Spaceflight Center and the home of the politically savvy vice president of the United States, Lyndon B. Johnson.

The Southwest had two other connections in Oklahomans James Webb and Senator Robert Kerr, chairman of the Committee on Aeronautical and Space Sciences. Interestingly, Webb had served as the assistant to the president in Kerr's own business, Kerr-McGee Oil. During Kerr's tenure, Oklahoma received its fair share of space contracts. Finally, after Representative Overton Brooks (D., La.) died in 1961, Representative George Miller (D., Calif.) became chairman of the House Committee on Science and Astronautics. Although Miller was a vocal critic of the expanding space program in 1961, he changed as chairman, becoming "more conservative, more forbearing, less inclined to criticize NASA and the space program."[77] Perhaps California's dominance in the aerospace industry, and its numerous potential contracts and jobs, helped quiet Miller.[78] In any case, the leadership of the space committees closely allied with the administration, helping to explain why NASA's programs and mushrooming budgets went virtually untouched during the crucial years of 1961 to 1963. From the beginning, the leadership intended to make the best possible case for rubber-stamping the program. They might have had a difficult time persuading committee members to go along with them had it not been for the flight of Soviet cosmonaut Yuri Gagarin.

Panicked by the Gagarin flight, members of the congressional space committees adopted the Kennedy administration's frontier narrative as a way to understand the space program and the Soviet challenge. During the uncertain days of the cold war, the familiar narrative reassured members that America had lost not its power but merely its way. In adopting Project Apollo and its depiction as a frontier narrative, members of congressional space committees reaffirmed America's mythic identity. Project Apollo went virtually unchallenged not only because members adopted the narrative as a way of understanding space events in the early 1960s but also because the committee leadership and the witnesses who testified at the hearings were self-interested and supported the administration's program.

Misleading Committee Reports

The criticism of Project Apollo, not drowned out by the committee members advocating the program, was relegated even further into the background in the final product of the space committees' deliberations: the reports. One can discover the space committees' strategy for "selling" the space program to Congress by examining these documents. Interestingly, the reports rarely mention the cold war arguments so prominent in the deliberations of the committee members. Instead, the space committees produced dispassionate, explanatory technical documents that ignored the rationales cited by committee members to justify the program. The space committees created reports to fit the expectations of a committee document, so as to give the impression that the hearings were thorough, technical investigations.

Both the House and Senate space committees created annual reports presenting the findings of their respective committees. The

House committee released its report first. In the three House reports from 1961 to 1963, one finds virtually no trace of the cold war arguments voiced in the various hearings. Only the 1963 report calls for a stronger emphasis on the "national security aspects" of the space program. The report, qualifying its recommendation, states that it was not suggesting any changes in the goals set out by the administration.[79] One can find perhaps the only instance of strong advocacy for Project Apollo in the 1961 report's description of *manned* space flight. "Exploration of space in its truest sense," the report asserts, "will begin only when man himself can participate directly." The report also cites man's unique ability to reason, to observe, and to deal with the unexpected. "Man," the report proclaims, "is destined to play a vital and direct role in the exploration of the moon and the planets."[80] The two-page description of Project Apollo stands out from the rest of the report, which is technical and informative. It is conspicuous because the committee lifted the material verbatim from a prepared statement submitted to it by Dr. Abe Silverstein, director of NASA's manned space flight programs, during authorization hearings that year.[81] The committee apparently abandoned the practice of printing large sections of NASA's material in 1962 and 1963.

The Senate's reports, released five to six weeks after the House report, were much shorter documents. The Senate committee often summarized parts of the House report, but much of the Senate's report followed its predecessor word for word. Thus not only did the two space committees present final reports that misrepresented their actual deliberations, they also presented members of Congress with virtually identical documents. The committees, it appears, wished to present a united front, thereby enhancing their credibility and that of their reports. Still another factor contributed to the ethos of the reports: they all received unanimous endorsement by the members of the committees.

Not all the committee members were completely happy with the reports or the authorization bills. In 1961 and 1963, congressmen added dissenting views to the text of the reports. In 1961, George Miller (D., Calif.), who became chairman of the House Committee on Science and Astronautics in 1962, and Perkins Bass (R., N.H.) blasted the House committee for authorizing $141,600,000 more than had been requested by the Kennedy administration. Complaining that no one had adequately explained or justified the additional expenditures, the two congressmen characterized the "precipitous manner" in which the full committee authorized the funds as "contradictory" to the thorough, intense, and specific examination characteristic of the subcommittees' investigations. The duo deplored the lack of "disinterested witnesses" who could offer "independent and objective evaluations" of statements made by industrial witnesses, whose statements Miller and Bass called "highly subjective" and "clearly motivated by self-interest."[82] Ironically, under Miller's tenure as chairman, the committee continued to rely on industrial witnesses with a vested interest in the program. He failed, however, to solicit the testimony of independent witnesses to evaluate the statements of representatives from industry.

In the 1963 report of the House Committee on Science and Astronautics, one finds additional dissent and the first clear sign of partisanship. Six Republicans—Richard L. Roudebush (Ind.), Thomas M. Pelly (Wash.), Donald Rumsfeld (Ill.), James D. Weaver (Pa.), Edward J. Gurney (Fla.), and John W. Wydler (N.Y.)—called for greater emphasis on military aspects of space exploration, particularly focusing on militarization of inner space—the areas of space surrounding the earth to a distance of 100 to 500 miles. Although characterizing the results of a manned lunar landing as "largely prestige," the group nonetheless stated that its members supported Project Apollo and the House authorization bill, since there existed "no other comparable program to develop space technologies at this time."[83] In a separate view, Representative Thomas Pelly questioned the urgency of Project Apollo, since the "technological and scientific benefits" were not commensurate with Apollo's $1 billion price tag for 1963. Thus he favored slowing the program.[84]

Undoubtedly, the 1963 report of the Senate Committee on Aeronautical and Space Sciences presented the greatest distortion. At the end, the Senate Committee attached a six-page summary of the ten scientists' testimony on America's space goals. The summary clearly attempts to distort the actual testimony in at least four ways. First, the report quotes only those passages that supported the administration's space program. The report, for example, characterized one of the two scientists who maintained that Project Apollo would provide military benefits, Dr. Lloyd Berkner, as an authority in the field of space and science, although the report offers no evidence for the claim.

Second, the report simply ignores testimony that directly challenged the scientific, economic, political, and defense aspects of America's space program. With regard to the defense issue, for example, the report ignores four scientists, including two Nobel laureates, who testified that the manned moon shot would make no contribution to America's national security. The report, moreover, fails to present the negative opinions of the most vocal critic: Dr. Philip Abelson. Only once does the report mention Abelson, noting his least critical comment—that NASA should make better use of its advisory groups. Nowhere in the report does it mention Abelson's contention that unmanned space exploration offered a viable, inexpensive alternative to manned space flight.

Third, the report trivializes the serious criticisms of the administration's space goals by understating them and by characterizing disagreement as typical of scientists. "As expected," the report explained, "the witnesses expressed different points of view on the questions under investigation." Finally, the report distorts the actual testimony by falsely concluding that the scientists agreed on the direction in which America's space program should go. "The hearings," the report proposes, "resulted in defining our space goals and explaining our motivations."[85] Citing only testimony that supported a manned lunar landing, ignoring testimony that directly challenged the administration's space program, trivializing what little disagreement it did acknowledge, and drawing false conclusions about the thrust of the hearings, the Senate space committee's report deliberately distorted the exact thing it supposedly set out to obtain: independent testimony that challenged government and industrial witnesses.

As NASA laid the foundation for its manned space program from 1961 to 1963, congressional space committees never seriously questioned the need for such a project. A host of factors explain why. Perhaps most important, space committee members, panicked by Yuri Gagarin's flight, appear to have accepted the Kennedy administration's frontier narrative as a way of understanding space-related events in the early 1960s. One should not underestimate the reassuring influence of this familiar narrative as a way of understanding and interpreting America's space program in the early 1960s. Respected scientists who testified before the Senate's space committee, for example, used the narrative to justify Project Apollo. And committee members did not question or chastise the scientists for doing so. One can also understand the lack of serious scrutiny given to Project Apollo in congressional space committees by considering the types of witnesses who testified before the committees and the composition and predilection of the committee leaders.

Committee members justified the program not only as a vital component of American national security but also as a race to regain national pride and international prestige. In addition, scrutiny of the committees reveals a striking instance in which scientists relied mainly on public and narrative argument, and not technical argument, when testifying on the scientific aspects of a program. This instance contradicts generally accepted notions of how scientists argue and further helps to explain why Congress offered little criticism of America's early manned space program. Finally, the space committees' reports offered a distorted view of their deliberations. The committees, attempting to produce a report that conformed to expectations—thereby enhancing the ethos of the documents and their committees' investigative abilities—presented dispassionate, explanatory reports that falsely portrayed the committee hearings as thorough technical investigations. One finds a striking example in the Senate's summary of the scientists' testimony of America's space goals. The summation distorted the hearings, omitting or trivializing the serious criticisms of well-respected scientists and emphasizing the testimony of a few scientists—with close ties to the administration—who supported the program without qualification. The committees used the reports to portray themselves as objective, technically competent, and highly credible investigative bodies. They hoped the ethos of the reports would help committee members peddle the program on the floor of Congress. In the final chapter of this study, we will examine how successfully proponents of the manned lunar program made their case in the debates over the space program on the floor of Congress.

6

Justificatory Rhetoric: Floor Debates Concerning Project Apollo

Members of Congress, writes John R. Fitzpatrick, make speeches on the floor for the "reading public," not for their colleagues. Speeches on the floor, Fitzpatrick observes, typically have "very little impact on the voting colleagues of the speaker."[1] Most genuine debate and decision making takes place not on the floor but in the various congressional committees. When bills come to the floor, members of the committees with responsibility for conducting hearings on the bills seek to justify the committee's recommendation, while other members often speak to justify decisions already made to the folks back home. In short, the rhetoric of congressional floor debates is typically not deliberative but justificatory rhetoric. Still, one can learn a great deal from studying justificatory rhetoric, because it allows one to better understand the forces that shape a given policy and discover the general sentiments and ideas of both speakers and audiences of a given time.

The congressional floor debates over NASA's manned lunar program clearly illustrated the justificatory character of the rhetoric of congressional floor debates. Seeking to entice the Congress as a whole into unquestioning acceptance of their recommendations, members of the space committee made the most of each opportunity to demonstrate their thoroughness in examining the program. They often commented on how many witnesses had testified before the committees, how many days they had worked in the committees, or the page length of the final committee hearings.

They also cited the integrity of the committee system as an argument. If one believed in the committee system, they argued, one should not question their final reports or recommendations. Furthermore, space committee leaders in the House never lost an opportunity to remind congressmen that the committee had approved its final report unanimously. Finally, some advocates argued that, ultimately, members must take the complicated and expensive budget on faith, putting trust in the leaders of the space program. In short, space supporters tried to rely upon the credibility of the committee process and the administration to sell the moon shot.

In addition, the arguments of space committee members on the floor of Congress emphasized not the technical or economic issues of concern to many members of Congress but rather the themes of hero worship and the desire for adventure. House space committee chairman George Miller (D., Calif.) opened the floor debates on NASA's FY 1963 authorization bill with a reference to the significance of beginning discussion of the bill "on the eve of our second orbital flight into space."[2] Not only did Miller highlight the more than coincidental timing of Scott Carpenter's flight, he also attempted to use the hero worship of the astronauts to justify NASA's big budget. Similarly, Miller and Representative Albert Thomas (D., Tex.), chairman of the Subcommittee on Appropriations, tried to use the astronauts to couch the debates in hero worship the following year. In the midst of floor debate in 1963, space critic H. R. Gross (R., Iowa) noticed the astronauts and their wives sitting in the House chamber and asked Thomas why they were present. Thomas explained that they were in town to receive an award at the White House and that he had invited them to his office. Suspicious, Gross pressed Thomas, asking whether it was "purely coincidental" that the astronauts came to the chamber just as congressmen were engaged in crucial debate over funding for the space program. Reluctantly, Chairman Miller rescued Thomas by admitting he had invited them to the House chamber. Miller explained that the astronauts were in town and, "like many other visitors in Washington, I presumed that they wanted to see Congress in session."[3]

Thus committee members succeeded in getting their colleagues in Congress virtually to rubber-stamp their recommendations in 1961 and 1962 without much debate over the specifics or merits of the program. James R. Kerr testifies to the lack of real floor debate on the authorization and appropriations of NASA's budgets in those two years, characterizing them as "routine and lifeless," with "almost a total lack of divergent viewpoints on the matter of getting to the moon."[4] One finds only a few individuals critical of NASA's budgets in these two years. In 1961, before taking over as chairman of the House Committee on Science and Astronautics the following year, George Miller blasted NASA's authorization bill. He complained that the committee recommended a "crash program." Pointing out that the committee started its hearings shortly after the Gagarin flight, Miller characterized the atmosphere as one of "panic" bordering on "hysteria." Miller concluded that he believed most of the committee would have approved "almost any figure on a crash program basis."[5] Yet by the end of the year, Miller became an ardent supporter. So little opposition arose in 1961 that the House passed the bill with a voice vote. And in 1962 the House passed NASA's authorization unanimously, 343–0. Representative H. R. Gross (R., Iowa) complained about the House's failure to engage in serious debate over such a massive spending measure: "I hope that if we do get to the moon we find a gold mine up there because we will certainly need it."[6] Nevertheless, Gross joined his colleagues in voting for the bill.

In 1963, one hears the first serious challenges to Project Apollo on the floor of Congress. For the first time, one witnesses serious and persuasive attacks on the scientific, military, economic, and educational justifications of the program. Yet, Congress still overwhelmingly approved $5.1 billion dollars for NASA; the House vote was a lopsided 335–47. Thus, even though critics raised serious challenges to some of the most important arguments in favor of the program, few seemed willing to go on record in opposition to the "great adventure."

This chapter examines the congressional floor debates over the early manned space program and reveals that while serious challenges to the scientific, military, economic, and educational justifications for Project Apollo arose during the debates, they were overwhelmed by the strains of the great frontier adventure story. Critics of the program either got caught up in the adventure story themselves or discovered the difficulty of refuting a story rooted deeply in America's mythology and its past. Whatever the explanation, the floor debates in Congress over sending a man to the moon became one last forum for celebrating the great frontier, and the result was a commitment to a *manned* space program that persists to the present day.

The Moon Shot and Congressional Critics

In 1963, serious criticism of Project Apollo arose within Congress for the first time. Two groups, one composed of social liberals, the other of military hawks, attacked the space program on a variety of diverse grounds, yet both groups called for a slowdown in the program's pace. Project Apollo met its greatest resistance in the Senate. On 10 May 1963 the staff of the Senate Republican Policy Committee released a report critical of the space program. It challenged the basic justifications for the moon shot and proposed that the government could better use the money to solve human problems on earth. The report's main thrust was that the program had proceeded too quickly. Calling for a slowdown in the attempt to reach the moon, the document proposed that the country would receive more value if "we take it easy and try to accomplish our ultimate purpose step by step."[7]

The loudest and most persistent criticism of the space program in the Senate came in 1963 during debate over NASA's appropriation bill. The critics, social liberals like J. William Fulbright (D., Ark.), Joseph S. Clark (D., Pa.), Maurine B. Neuberger (D., Ore.), and Ernest Gruening (D., Alaska) wanted to slow the program's pace, thereby saving money that could be used to solve social ills. Fulbright led the charge. He questioned the administration's "excessive emphasis on space in relation to other national programs," most notably in education and employment.[8] Like Fulbright, Senator Ernest Gruening viewed solving the unemployment problem as crucial to the nation. He was "far more concerned" with the 5 million unemployed in the country than with landing "men on the moon at the earliest possible date."[9] Joseph Clark, a critic, tried to dramatize the enormous cost of the lunar program by arguing that for the cost of one of Project Apollo's "Cadillac sized modules," the government could rebuild "a fair sized city from the ground up."[10]

In 1963, Representative Louis C. Wyman (R., N.H.) and William Fulbright offered amendments to cut NASA's appropriations. Wyman proposed a $700 million reduction, cutting $550 million from Project Apollo and an additional $150 million from lunar and planetary exploration. He explained the objective of his amendment in the following way: "It will defer, it will stretch out, it will delay a moonshot program and this we should do." He called the current schedule a "crash program."[11] The bill gained some support in the House but lost 132–47.

Fulbright, the most vocal critic in 1963, took a similar position. In offering an amendment that would cut 10 percent across the board from research and development, construction of facilities, and administrative costs, Fulbright explained that his amendment attempted to "allow time to re-evaluate the goal of trying to reach the moon in this decade and to proceed on a more deliberate and thoughtful basis." He did not object to the moon shot, he insisted. "I merely object," Fulbright announced, "to our trying to go tomorrow."[12] The amendment lost in the Senate in a surprisingly close vote, 46–36.

The first substantive debate over Project Apollo centered around questions of political prestige and national defense. A poll of House members in early 1962 found that many believed the space program "questionable except when viewed from the standpoint of the cold war,"[13] suggesting the success of administration arguments on the issue. Yet the administration had a lot of help from the congressional committees on the floor. Representative Albert Brooks (D., La.), for example, chairman of the House Committee on Science and Astronautics until his death in late 1961, opened the NASA authorization debate that year by citing the importance of the moon shot for America's prestige around the world. Brooks called the space program one of America's "strongest weapons" in the "ideological struggle with communism for the mind of man."[14] Similarly, Representative J. Edward Roush (D., Ind.) cited checking communist expansion as justification for supporting a lunar landing. The moon shot, Roush declared, gave America a chance to "go on the offensive in a peaceful way in this great battle in the cold war."[15] Perhaps Representative H. R. Gross (R., Iowa) best summed up the opinion of many members: "In this cold war fight with the Communists, the new battlefield is space."[16] Thus the moon shot came to represent a test of national vitality. Losing it would greatly reduce America's prestige and influence overseas. Representative David S. King (D., Utah) characterized outer space as a "showcase" in which the world would judge the United States, while Senator Clinton Anderson (D., N.M.), who became chairman of the Aeronautical and Space Sciences Committee in 1963, argued that losing the race to the moon with the Soviets would be "disastrous" for American prestige abroad.[17] The administration's most reliable advocate, Representative Olin E. Teague (D., Tex.), pointed to the "uncommitted" nations and what effect losing the space race would have on them. "If we flunk the space test," Teague concluded, "our prestige will dwindle away to nothing."[18]

Critics of the moon shot, however, were not about to concede its symbolic value in the cold war. Perennial advocate of increased social programs, Senator William Fulbright attacked the lunar landing program on the grounds that "not a single nation" had succumbed to Soviet influence "because of communist successes in space." The prestige garnered from landing on the moon would be "fleeting," he insisted, and he characterized the moon mission as a "9-day wonder of history, a gaudy sideshow in the real work of the world."[19] Only a small group of his colleagues seemed to share Fulbright's opinions. Senator Joseph Clark was among the few agreeing that there was an "immature" attitude behind the space race; he compared it to a "high-school cross-country race."[20]

Other critics alleged that partisan politics, not national prestige, best explained President Kennedy's support for a moon shot. Representative Thomas M. Pelly (R., Wash.) suggested that Kennedy was more worried about his own prestige "since he took office." Pelly called the moon shot a "costly political stunt."[21] Representative James D. Weaver (R., Pa.) likewise blasted the narrow political motivations of the president, insisting that the Kennedy administration challenged the Soviets to a lunar race to divert public attention away from "the fiasco of the Bay of Pigs."[22] Kennedy, moreover, left himself and NASA open for attack about the sincerity of their earlier cold war arguments when he proposed a joint U.S.-U.S.S.R. moon mission. A joint venture would eliminate the competitive aspects of the program. Worse yet, it would also eliminate another of the advocates' main justification for a manned moon mission: national defense.

The greatest challenge to Project Apollo came in response to arguments that it would enhance the defense of the country. One can divide the members of Congress into three camps on the defense issue surrounding the program: those who supported the defense value of a moon shot, those who questioned the defense value of a moon shot and wanted to scale back the space program, and those who questioned the value of a moon shot and wanted greater emphasis placed on a military space program.

Space committee members led those defending the contribution of Project Apollo to the country's national security. Olin Teague, chairman of the House Subcommittee on Manned Space Flight, cited national security as the "most important reason for going to the moon." If America's defense did not "depend" upon a moon shot, Teague added, he would not favor the program as much as he did.[23] When critics began challenging the moon shot in 1963, Teague impatiently insisted that more than "90 percent" of the House members "know" Project Apollo enhances national security, and "that is why most of them support it."[24] Space committee members, such as Senator Stuart Symington (D., Mo.) and Representatives George Miller (D., Calif.), James G. Fulton (R., Pa.), Albert Brooks (D., La.), and Emilio Daddario (D., Conn.), joined Teague in justifying Project Apollo in the same terms. They argued that America could not let the Soviets, or any other hostile power, control outer space. Doing so would jeopardize America's security.[25]

A small but vocal group in 1963 began to challenge the military value of going to the moon. Senators William Fulbright and Joseph Clark, joined by the eternal congressional gadfly Senator William

Proxmire (D., Wis.), questioned the contribution that Project Apollo made to the country's security. These critics came well prepared. In addition to marshaling their own evidence from respected sources, they also cited criticisms voiced in the Senate space committee hearings but ignored in the committee's 1963 reports. Critics relied heavily on the concerns of scientists like Dr. Harold Urey and Dr. Polykarp Kush, two of the ten experts who appeared before the Senate Committee on Aeronautical and Space Sciences. During debate over NASA's appropriations in 1962 and 1963, Fulbright and Proxmire consistently challenged administration charges that the moon shot would help America's defense. In a 1962 speech, Proxmire pointed out the superiority of land-based missiles over space or moon-based missiles. Firing an object "moving at such fantastic speed in space, or even to fire from the moon," he noted, "would result in all kinds of handicaps, all kinds of limitations" not imposed on delivering the same payload from earth.[26] Proxmire cited numerous experts to support his position: Dr. James R. Killian, president, Massachusetts Institute of Technology and former science adviser to Eisenhower; Dr. Lee A. DuBridge, president, California Institute of Technology; and Dr. Harold Brown, director, Defense Research and Engineering. Not only did these men question the value of space weapons, they also questioned the military need for manned space vehicles. Proxmire went so far as to say that space spending could have an "adverse effect" on American national security.[27]

Senator William Fulbright joined Proxmire in attacking the military rationale for the program. In a speech delivered on 19 November 1963 Fulbright called the military justification for a moon shot "minimal." Fulbright cited General Maxwell Taylor, chairman of the Joint Chiefs of Staff, and Nobel laureate Dr. Harold Urey, who said the moon shot would have no direct bearing on America's military security. Furthermore, he introduced the testimony of General Curtis LeMay, chief of staff of the air force, who viewed ground weapons as the most effective and least costly. Characterizing the manned moon mission as "not essential to the Nation's security," Fulbright, like Proxmire, called for a reduction in space spending. Senator Stuart Symington, a space booster, attempted to counter Fulbright's charges, arguing that the people in the military whom he respected the most understood the "vital importance" of space to national defense. Symington noted that the military had never been known for its "unanimous decisions" and insisted that it was not uncommon for the three military branches to have three entirely different ideas about a military issue.[28] This argument also proved handy in answering the last group of critics the administration faced—those who agreed that Project Apollo made a minimal contribution to national defense but saw great defense potential in a different sort of program.

Since NASA's inception, a small group in Congress began calling for a greater emphasis on a military program to counter the Soviets' growing military space program. Quintessential cold warrior Senator Barry Goldwater (R., Ariz.), joined by Senator Howard W. Cannon (D., Nev.), demanded a greater emphasis on military uses of space, cautioning that the Soviets did not make a distinction between military and civilian programs. George Miller, chairman of the House space committee, defended the administration's position, declaring that "our defense officials are not dolts." Miller echoed Symington's argument that the "real cause" of the "squabble about the military in space" was an "inhouse difference of opinion in the Military Establishment."[29] The movement for a greater military emphasis gained momentum in 1963 as six Republican congressmen attached a minority report to the report of the House Committee on Science and Astronautics calling for greater emphasis on military exploitation of inner space.[30] During floor debates, additional members joined in pushing for a greater emphasis on inner space.[31] James Weaver's comments typify those of the critics. America can only wage the "cold war" with the "Communists," Weaver warned, "if our space program progresses with national security as its prime goal."[32] Frustrated with lack of support for his position, Weaver demanded that NASA administrator James Webb either realign America's space "objective" to national security or resign. Viewing the moon shot as a public relations gambit, Weaver declared that "Congress and the public can no longer tolerate public relations gimmicks and doubletalk concerning the space program and our space gap when our national security is threatened."[33] These critics wanted national security as the first priority of America's space program. Representative Donald Rumsfeld of Illinois stated their desire clearly: "This country should direct itself toward inner space and not place our top priority in the direction of the moon."[34]

One cannot overestimate the significance that members of Congress placed on the cold war and defense arguments in debate on the floor of Congress. The House perhaps best demonstrated its distrust of the Soviets and the depth of its cold war sentiments by passing an amendment prohibiting the administration from entering into a joint space effort without the consent of Congress. The amendment, offered by Thomas Pelly, won a narrow victory, 125–110, but it was one of the few times that the administration's space policy was flatly rejected in Congress.

Space committee members also had to defend the administration's position on the floor of Congress against attacks on the scientific value of the manned lunar program. On the floor, the scientific arguments in favor of the program were much the same as in the committees. Representative Emilio Daddario (D., Conn.) argued that the space program would contribute greatly to America's "knowledge of the basic sciences."[35] Likewise, Senator Margaret Chase Smith (R., Maine), whose letter to Chairman Clinton Anderson prompted the committee hearings of the ten distinguished scientists in June 1963, proposed that the scientific fallout from the moon shot might surprise the country. The "valuable scientific knowledge that we gain on the way to putting a man on the moon," she conjectured, "could very well be far more significant than the end objective."[36] Representative James Fulton even predicted that the program would create "urgent demands for more knowledge from science."[37] Finally, George Miller, a onetime critic of the program, relied heavily on the scientific appeal, especially when the program started coming under attack in 1962. On three different occasions in 1962, Miller felt it necessary to tell fellow congressmen that the space flights were not shows or stunts but "serious scientific experiments."[38]

Armed with testimony from many of the scientists who testified before the space committees, critics raised serious challenges to the scientific value of Project Apollo. Again, one finds Fulbright, Proxmire, Cannon, and Thomas Pelly (R., Wash.) at the forefront of the attack. Pelly, for example, questioned the scientific contribution of a manned lunar landing. "A high percentage of the scientific fraternity," Pelly proclaimed, "find fault with the Apollo program on many specific scores." He went on to cite numerous scientists, among them Dr. Philip Abelson and Nobel laureates Dr. Harold Urey and Dr. Polykarp Kush, who refuted Project Apollo's scientific contribution. Pelly concluded by advocating that the United States send instruments to the moon instead of a man, since they cost less and would gather the same amount of scientific information.[39] Similarly, Fulbright used the testimony of distinguished scientists to challenge the scientific justifications of a moon shot. No scientists, he declared, justify Project Apollo's "pace" or "cost as essential to scientific objectives."[40]

Proxmire, supported by Cannon, went beyond merely questioning the scientific objectives of the program to argue that Project Apollo might actually hurt American science, education, and industry by stealing scientists away from those areas. America, he submitted, already suffered from a shortage of scientific manpower. Project Apollo would only exacerbate the problem. In mid-1962, Proxmire offered an amendment to NASA's appropriations bill establishing a commission to study the impact of NASA's programs on scientific manpower in the United States. Responding on behalf of the space committees, Senator Robert S. Kerr (D., Okla.) insisted that Proxmire had greatly overstated the shortage and that the president already had a commission overseeing the question of scientific manpower. The Senate rejected Proxmire's amendment 83–12.

Space boosters ultimately tried to refute the scientific criticisms of the administration's space policies by raising questions of expertise and credibility. Ken Hechler (D., W.Va.), for example, a member of the House space committee, submitted that the congressmen raising scientific objections were not "scientists" and therefore could not make an informed opinion on many aspects of the program. "Let our experts run this program," Hechler implored. "Let us not substitute our layman's judgment for theirs."[41] Chairman Clinton Anderson attempted to denigrate the testimony of many of the scientists themselves, characterizing it as motivated by self-interest. Anderson called the disagreement among the scientific community "an understandable expression of parochial interest." Many scientists critical of the space program were upset because they wanted their disciplines to receive funding. "Any scientist worth his salt," Anderson explained, "should be an advocate for his own discipline."[42] Thus, in one full swoop, Anderson tried to make all scientific criticism suspect. Interestingly, by Anderson's logic, one would have to assume that scientists who advocated the program also did so out of "parochial interest."

Supporters of the administration's space program encountered similar opposition when they argued that Project Apollo would stimulate the economy. Appealing to the average American by arguing that Project Apollo would create new jobs and stimulate economic growth, Olin Teague characterized the program as "stimulating employment to a degree little recognized by most people." Teague claimed that it created "millions" of jobs if one included those in education and industry along with jobs related directly to NASA.[43] Other advocates emphasized how the moon shot would create new consumer goods and whole new industries. Albert Brooks warned that America's "economic and material well-being" depended "in no small degree" on what the country accomplished in the space program.[44] Olin Teague metaphorically summarized the space committee's view on the economic impact of the program during the 1963 debates over NASA's authorization bill. The space program "started the blood coursing a little more fervently through the arteries of our economy," Teague stated. It started the industry's "pulse" "beating like a drum." Teague concluded by predicting that the space program would trigger a "new industrial revolution."[45]

Critics of the administration's space program questioned the impact of the moon shot on the civilian economy. William Fulbright charged that the space program might actually "become a drain on the civilian economy," jeopardizing America's position in "world trade." He countered the notion that the moon shot would stimulate industry by pointing out that American "aerospace industries," where the country spent its billions of space dollars, did not need "stimulation."[46] Representative Louis Wyman (R., N.H.) attacked the relationship between NASA and big business, arguing that it had become too cozy. Stating that certain businesses had a vested interest in the space program's growth, Wyman insisted that a number of the contracting companies had become "propagandists for NASA and the moonshot."[47] Not surprisingly, the most vocal economic critic was William Proxmire. He agreed with space advocates that the program would stimulate the economy. But, he noted, it would be "bought at a stiff price in higher taxes." In 1962, Proxmire attacked NASA for its lack of competitive bidding. Warning that the lack of bidding harmed small business and led to bigger contracts and greater concentration among the few big businesses who received space contracts, Proxmire offered an amendment to the Senate appropriations bill that would force NASA to adopt more competitive bidding for its contracts.[48] Administration space supporters Robert Kerr and Stuart Symington responded to Proxmire's charges, arguing that the amendment would cause unnecessary red tape and that only big businesses had the money necessary for the research, design, and manufacturing needed for the new space equipment. The amendment gained some support but lost 72–23.

Finally, space boosters argued on the floor of Congress that the manned lunar program would have a positive impact on education in America. James Fulton advanced the education argument in mid-1963, proposing that the space program forced American universities to demand "higher scholastic achievement through greatly improved curriculums" so that its graduates would be useful to NASA, the DoD, government, and to "society as a whole."[49] Super space booster Olin Teague concurred, calling the space program a "gigantic spur to our educational system."[50] Fulbright and Proxmire challenged the educational argument, warning that the program would only drain much needed science teachers away from the universities.[51] The educational argument undoubtedly had great popular appeal, testifying once again to the concern in the floor debates with public justification.

Critics of the administration's manned lunar space program clearly offered serious and legitimate arguments on the merit of the program. Always armed with expert testimony, critics argued forcefully against the scientific, military, economic, and educational justifications of a manned lunar landing. Despite these objections, Congress continued to provide overwhelming support for Project Apollo, appropriating billions for the manned lunar effort from 1961 to 1963. How does one account for this? Why did the critics have so little influence? The answer may lie in the ultimate justification for the manned lunar landing during congressional floor debates: the promise of a great frontier adventure. As in the rhetoric of the Kennedy administration and the popular mass media, the frontier adventure story seemed to transcend and overwhelm debate on the merits of the Kennedy administration's plan.

The Frontier Narrative: Irresistible and Irrefutable

Members of Congress, no less than administration officials, romantically portrayed the space program as a glorious frontier adventure, reminiscent of American adventures of the past. Many members may have invoked the comparisons to reassure themselves that traditional American values and know-how would work in meeting the challenges of the future. Perhaps they hoped to restore their own faith, and the public's, in the efficacy of the frontier myth. Whatever the explanation, Olin Teague apparently expressed the feelings of many in Congress when he observed that before the space race, America believed there were no more first-class "challenges," no more "new frontiers." The idea of lunar exploration, he concluded, had "reawakened" America's "spirit of adventure and achievement" like nothing since "the days of the pioneers."[52]

Echoing administration advocates, space boosters in Congress placed space flights within a long tradition of frontier exploration. Clinton Anderson (D., N.M.), for example, declared, "As surely as the ancient mariners ventured across unmapped seas, as certainly as explorers opened up our West, and men established bases at the barren and inhospitable poles, we are going to continue this bold journey through space."[53] Other space committee members voiced the common theme that a pioneering spirit was endemic to the American people. Representative R. Walter Riehlman (R., N.Y.) announced, "We are a pioneering people, a pioneering nation,"[54] while Senator John J. Sparkman (D., Ala.) attributed America's rise to world leadership to "a nature which drives us to explore the unknown."[55] And much as politicians in the past had argued that the western frontier would alleviate the crowded conditions on the eastern seaboard, Olin Teague characterized space as an "outlet" for a civilization fast becoming "dangerously cramped and limited by its confines to this planet."[56]

In debates over Project Apollo on the floor of Congress, one finds all the basic constituents of the traditional frontier adventure story: (1) an identifiable, conquerable geographic location that is (2) unknown and hostile and includes (3) a malevolent antagonist who is thwarted by (4) a heroic adventurer. Members of Congress, however, placed their greatest emphasis on the heroic adventurers and their flights.

Describing the flight of Alan Shepard, Representative Chester E. Merrow (R., N.H.) said that in "blazing the pathway to space," Shepard had become a true "pioneer."[57] Similarly, Representative Thomas J. Lane (D., Mass.) submitted that Shepard carried forward an "old but ever-young American tradition; to part the curtains leading to the unknown, in man's searching for a greater understanding of himself and his total environment."[58] In glorifying and reaffirming the traditional American spirit, members of Congress reassured themselves and the country that American institutions and the American people could, as in the past, meet the challenges before them. One can see this feeling reflected in the statement of Karl E. Mundt (R., S.D.). Calling Shepard's accomplishment more important than merely a scientific or technical achievement, he emphasized instead how it "once again demonstrated the strength and character of our people."[59] Numerous speakers also commented on Shepard's personal qualities, which they were quick to interpret as traditionally American. During authorization hearings months after the flight, space advocates like Olin Teague were careful to stress the importance of Shepard *the man* to the success of the missions.[60]

Not surprisingly, the most celebratory rhetoric came in response to the flight of John Glenn. The members described Glenn within a narrative framework as the quintessential American pioneer engaged in the quintessential American adventure. Speakers relied heavily on comparisons with traditional American adventurers when praising Glenn and his feat. Senator Alan Bible (D., Nev.) insisted that Glenn's "stirring adventure into the unknown bore with it the spirit of pioneer Americans before him who met the challenges of their times."[61] Representative Harold D. Donohue (D., Mass.) even described Glenn's wife and family as traditional pioneers. The Glenns, Donohue announced, consistently exemplify the "traditional pioneering spirit upon which this great Nation was founded and which will sustain it forever against any and every challenge of the future."[62] Perhaps Carl Albert (D., Okla.), future Speaker of the House, made the most sweeping comparison. In Glenn's feat, Albert submitted, there was "something of Columbus when he crossed the Atlantic. There was something of the Pilgrim who faced the dangers of the wilderness and the Red man. There was something of the daring and devotion of the men who followed Washington at Valley Forge. There was something of the pioneer who turned back the frontier and built this Republic." Albert concluded with perhaps the most striking overstatement of the entire debate when he called Glenn's flight "one of the most heroic efforts in this history of man."[63]

Perhaps as important as Glenn's feat were the personal qualities attributed to him during the floor debates. Representative Fred Schwengel (R., Iowa) observed that the more he heard about Glenn's flight, the more "impressed" he became with Glenn's "sterling qualities as an individual."[64] Senator Ralph Yarborough (D., Tex.) described him as a "living symbol," while Senator Mike Mansfield (D., Mont.) said that he "exemplified the type of American which many thought was disappearing from the scene."[65] Similarly, Representative John M. Butler (R., Md.) called Glenn the "epitome of the normal virtues," like reliability, diligence, friendliness, and loyalty. Butler added that in Glenn, those virtues had been "triumphantly heightened, purified, and unified."[66] Glenn epitomized the rugged frontiersman of the past. He epitomized traditional American values, which accounted for his personal success and had also made the country great.

Space advocates leaned heavily as well on more general themes of the frontier motif in justifying the space program. Robert Kerr, for example, maintained that the space program would "enable the American people to meet their destiny."[67] Similarly, Representative Richard H. Fulton (D., Tenn.) placed NASA administrators and astronauts within a long line of heroic pioneers. Some people ask why America should go to the moon, Fulton observed. "The drive to explore the unknown led to the building of our great nation, it led to its discovery." Many people, Fulton recalled, viewed Columbus as "insane." Hudson, De Soto, Crockett, Boone, Lewis and Clark, all supposedly madmen, followed after Columbus, blazing "a trail" for others to follow. Today America has a new breed of "nuts," touched by "lunar madness": the "Gilruths," the "Webbs," the "Gordon Coopers," and the "John Glenns." These men, Fulton predicted, "will lead a great America in the conquest of outer space." Fulton concluded by saying he supported increased funding for Project Apollo "because it is in keeping with the pioneer spirit of this great Nation."[68]

The narrative's emphasis on the "frontiersman" was not lost on the members of Congress. They used the exciting, romantic flights to justify the need for *manned* exploration and to denigrate machines. Immediately after Shepard's flight, Representative Thomas Lane proposed that America believed "man must be the master of this technological progress, not its robot slave."[69] Glenn's flight also inspired such comments. Edward Roush maintained that Glenn's flight demonstrated the "great value of having human control." Calling the "human component" the most important part of the mechanism, Roush insisted that one could not "overemphasize" the "human element" in America's space program.[70] One also finds glorification of the individual in the comment of Representative Garner E. Shriver (R., Kans.) about Gordon Cooper's flight. "When the automatic electronic devices went inoperative at the crucial period, it was the man in the capsule who manually directed" it to a perfect landing in the Pacific.[71] The frequent use of the frontier mythology and the denigration of the machine in the floor debates reflected an almost unquestioning acceptance among members of Congress of the need for *manned* space exploration.

Significantly, even critics of the program seemed to be caught up in the romance of the great frontier adventure story. Congressional penny-pincher William Proxmire, observing that "we are on the brink of this adventure into space," finally conceded in 1962 that "we have to accept the costs involved."[72] Senator Gordon Allott (R., Colo.), another frequent critic of the program, also succumbed to the frontier mythology when describing fellow Coloradan Scott Carpenter. The astronaut, Allott observed, eagerly anticipated his opportunity to become "one of this country's pioneers into space, just as his forbears were pioneers into another new world about which little was known just a few generations ago, the Western frontier."[73] Perhaps the most surprising statement, however, came from space critic Ben F. Jensen, a Republican congressman from Iowa. Extremely critical of Kennedy's supplemental bill before finally supporting it in 1961, Jensen marveled at how John Glenn had "challenged the elements of outer space and returned safely back to earth." After watching the reaction of the crowd during a parade for Glenn, Jensen proclaimed that he now felt "more satisfied than ever" that he had acted correctly in supporting additional funds for the space program during the "last session of Congress."[74]

A few critics attempted to challenge the hero worship of the astronauts and the romantic portrayal of the moon shot. Representative John Lindsay (R., N.Y.) was among a small group of congressmen who had apparently not become moonstruck. After listening to James G. Fulton (R., Pa.) praise Glenn as a true role model for youth of America, Lindsay responded that no one could "disagree" with Fulton's comments about the astronauts' "stature and rectitude" or the desirability of having America's teenagers "uplifted" by the astronauts' feats. "But," Lindsay noted, "we would be enlightened if we could hear some discussion of the bill, what is in it, what for, where we are going, and the timetable."[75] Senator Maurine Neuberger (D., Ore.) likewise challenged the frontier motif when criticizing the program. She reminded the Senate that in the past, Americans "conquered many new frontiers." But, she argued, America should first take care of the needy here on earth before endorsing a "great romantic—and extravagant—adventure."[76] William Fulbright, however, sounded the most direct criticism of the frontier adventure story of prospace advocates. The moon shot, he argued, was not "essential" simply because it was "new and creative and adventurous." Challenging the comparisons between a moon shot and the voyage of Columbus, he called the analogy "inaccurate and oversimplified," since private sources financed a good deal of Columbus's mission. Moreover, he concluded, Columbus's conquest brought Spain "only a brief period of glory" followed closely by "four centuries of political and economic decay."[77]

Few of Fulbright's colleagues seemed moved by his critique of the frontier adventure story. Perhaps frontier mythology is simply too deeply embedded in our culture to be "refuted." Or perhaps the narrative form of the adventure story protected it against refutation by "logical" argument. As William Lewis has suggested, narratives are not refuted by arguments. Rather they must be *supplanted* by alternative narratives that ring truer for the audience.[78]

With the cold war in full swing, Americans wanted desperately to have faith in a viable narrative that held out hope for America's future. Whether "realistic" or not, the frontier adventure story was appealing to Americans. And members of Congress were not immune to its influence. Americans *wanted* to believe the myth of the frontier adventure when faced with the insecurities of the cold war in the early 1960s. Congressional floor debates over America's manned space program from 1961 to 1963 reveal that members of Congress supported Project Apollo primarily as a result of narrative appeals.

Conclusion

The space program stood as the centerpiece of the Kennedy administration's New Frontier. It symbolized the optimism, the excitement, and the advancing science and technology of the future. It allowed Kennedy to make good on his promise to get the country moving again. At the same time, it called forth traditional American values—the sacrifice, hard work, and self-determination of the Puritans and the courage, ingenuity, and rugged individualism of the pioneers. The program, in short, became a symbolic amalgamation of the old and the new.

One theme addressed in this book is the contrasting nature of technical and narrative arguments and how those arguments may play different roles in public discussion of social policy. The Kennedy administration's public relations campaign on behalf of Project Apollo established a number of rationales for a manned lunar landing. Led by President Kennedy and elaborated by NASA's own public relations office, the campaign highlighted a number of political, scientific, military, and economic arguments for a manned lunar landing. Ultimately the Kennedy administration's rhetoric, reinforced by the media, emphasized not technical arguments but a narrative with deep roots in American history and culture. Depicting Project Apollo as a great frontier adventure story, the administration enticed the public and Congress into thinking of a moon shot not in "logical" terms but as a reaffirmation of a romantic American myth. Moreover, the myth could not easily be refuted by technical claims grounded in a more "rationalistic" worldview.

In public discussion of social policy, technical and narrative arguments seem to play different roles. Technical argument, although crucial to the development of a great deal of social policy, limits public discussion to experts in a given field and excludes the public and most members of Congress. In contrast, narrative argument allows expert and nonexpert alike to participate. Moreover, narrative argument may be crucial in building and sustaining public understanding and support for social policy, since the public may never fully understand the technical arguments of the experts and may therefore not be willing to support some social policy. Finally, narrative appeals appear difficult to refute with technical argument. In some instances, narratives may subsume technical argument, leading to misguided, wasteful, or even disastrous social policy.

A second theme addressed in this book is how both the executive branch and the news media function to help set the agenda in American politics. The story of space as the New Frontier was tailor-made for the popular print media of the day. While prevailing journalistic values led a number of magazines and newspapers to criticize specific aspects of the program, few could resist joining the celebration of a manned lunar expedition as a great frontier adventure. The most noteworthy relationship between the government and the media, of course, was *Life* magazine's exclusive contract with the astronauts. Telling the astronauts' story in striking pictures and patriotic prose, *Life* went well beyond cheerleading to become little more than an arm of NASA's public relations.

Even Congress seemed to be swept away by the rhetoric of space as the New Frontier. The members of congressional space committees not only failed seriously to question the need or feasibility of a manned lunar landing but also actively sought ways to justify the program. Subsequently pleading the administration's case during floor debates over the manned space program, committee members often found it difficult to refute the military, scientific, economic, and educational objections voiced by their colleagues. Even the critics fell under the spell of the great frontier adventure story. Either influenced by or quite sensitive to the power of the frontier myth in American public consciousness, some of the most skeptical members of Congress ultimately echoed its refrains.

In describing space as a New Frontier, the Kennedy administration shaped the way in which Americans interpreted and gave meaning to space exploration for many years to come. Deeply rooted in America's cultural mythology, the frontier narrative subsumed arguments about the technology and economics of the program, and it established a presumption in favor of massive commitments of the nation's resources to staffed space flight. Prior to John Glenn's flight in February 1962, Arthur C. Clarke wrote that NASA was creating "the myths of the future at Cape Canaveral."[1] The present book has emphasized how the rhetoric of the Kennedy administration *recreated* the myths of America's past. Clarke's observation remains instructive. As the space program has evolved, the frontier myth has apparently continued to carry as much, if not more, might than technical considerations. The continuing influence of the frontier mythology is perhaps nowhere more evident than in the decision to develop the space shuttle program.

Scientific and commercial research has always played a "secondary role" in the space shuttle's missions.[2] From the start, NASA justified the space shuttle primarily as a cheap, safe alternative to unmanned launchings of satellites. Yet unmanned, expendable rockets could put a satellite directly into geosynchronous orbit, while the shuttle orbited too low to accomplish such a task. Thus the shuttle had to resort to carrying its satellites into space and then releasing them, where they then used their own boosters to blast themselves into orbit. NASA also insisted that human beings were essential to its space flights because human beings could pluck satellites from orbit and return them to earth for repairs. The shuttle could rescue only the lowest orbiting satellites, since it could reach only 500 miles above the earth, while many of the most important satellites orbited over 22,000 miles up.

Since the *Challenger* disaster, moreover, the cost of launching satellites via the shuttle has "skyrocketed" to more than double its preaccident cost. If the shuttle's primary purpose is to launch satellites, and if unmanned rockets can do the job better at approximately half the cost, then there seems little technical or scientific rationale for continuing the program. Now, more than ever, as William Boot suggests, the shuttle can "do little, if anything, of importance in space that machines could not do better."[3]

Part of the explanation for the continuing commitment of vast resources for the shuttle program may lie in its "fit" with frontier mythology. One should not overlook the symbolic importance of the shuttle's ability to carry crews of human beings—and of the ability of those crews to fly the shuttle like an airplane. Although the astronauts still actually *do* little to control their destiny aboard the shuttle, their presence and their ability to land the spacecraft provide at least some reaffirmation of the daring and rugged individualism of the frontier myth.

At the same time, an important element of the frontier story faded as space shuttle launches became routine: a concrete destination or a frontier to conquer. With no tangible destination to lend excitement to the flights, NASA tried for a time a series of "gimmicks" to generate interest: night launches, teachers and journalists in space, and so on. With no new destinations in space, the shuttle program became a frontier adventure story without the frontier. Shuttle flights were flights to nowhere, with nothing to conquer.

President George Bush tried to recapture the power of the frontier narrative by identifying clear objectives for the staffed space flight program, thereby making the narrative structurally coherent once again. On 20 July 1989, the twentieth anniversary of America's landing of a man on the moon, Bush established three objectives for the space program: developing space station *Freedom*, returning to the moon, and sending a "manned" mission to Mars.[4] Undoubtedly, Bush and NASA hoped to use the Mars mission to recapture the frontier spirit of the lunar landing. They realized just how crucial the goal of "conquering" the moon was in sustaining the mood of frontier adventurism during the early space program.

Just as Bush identified clear objectives that reestablished the narrative's structural coherence, the Soviet Union dissolved, creating a new problem for supporters of staffed space flight. Without the Soviet antagonist in the frontier narrative, the story may once again lose its structural coherence and, consequently, its persuasive power. If President Bill Clinton wishes to support a vigorous, staffed space flight program, it appears that he will have to identify a new antagonist in space to make the frontier narrative "ring true" in the 1990s. Without a plausible antagonist in the frontier narrative, the staffed space flight program may fail to create the excitement and support it will need to continue as envisioned by Bush and NASA.

If President Clinton, or his successor, is able to restore the frontier narrative's structural coherence, he or she will find the press ready to support a staffed mission to Mars. One senses in the media an eagerness to jump on the mission-to-Mars bandwagon. In an article in *U.S. News and World Report*, William J. Cook sounds much like the journalists of an earlier era as he describes the need for new "adventures" to enrich "the human spirit."[5] Although Cook acknowledges that one cannot justify sending human beings heavenward economically, scientifically, or militarily and that robotic probes would be more productive scientifically than manned missions, he observes that "more than bread feeds the human soul" and calls a manned mission to Mars "the one space adventure that most fires the Earthbound imagination."[6] A mission to Mars also has a strong advocate in popular astronomer Carl Sagan, who has aired his plea for exploration of Mars in various forums.[7]

Perhaps a mission to Mars will be just the ticket to recapture the frontier spirit in the American space program. Meanwhile, however, American space exploration has fallen behind that of the Soviets in many aspects of staffed space flight in the past twenty years. In addition, the French, the Japanese, and even the Chinese have entered the field of space exploration, emphasizing the scientific and technical missions slighted in the once all-powerful American program. The frontier mythology thus creates a dilemma unique to the American context. On the one hand, restoration of the frontier spirit may be necessary to sustain popular and political support for the space program—and, of course, to win support for NASA's ever-growing budget. On the other hand, the frontier myth might continue to steer American space policy toward manned space programs that sustain the myth but are technically and scientifically inferior to the programs of competing nations.

The final theme that the study has addressed is the problematic relationship between technology and human choice. America's space program, with its emphasis on staffed space flight, "controls" or limits the kind of exploration in which it can engage, the kinds of experiments it can conduct, and, ultimately, the kind of information and knowledge it can gain. The book suggests an irony of technology: the more technology human beings create to control their environment, the more the technology shapes the environment and hence "controls" or limits human choice and action.

Whatever the impact of frontier mythology on the future of the space program, another and perhaps more far-reaching legacy of the early space program remains: a faith in the ability of the federal government to marshal the economic and technical resources of the nation to solve virtually any problem.

As Walter McDougall points out, the Kennedy administration's space program led to an unquestioning acceptance of the institutionalization of technological change for state purposes. In answering the challenges of a race into space with the Soviets, Americans also accepted a greater concentration of governmental power. Kennedy's call for an expanded space program, McDougall writes, amounted to "a plea that Americans, while retaining their free institutions, bow to a far more pervasive mobilization by government, in the name of progress."[8] Americans, while cheering the individualistic spirit of the astronauts, also came to a greater acceptance of an activist government. Americans seemed to believe that in the country's vastness, the individual and the burgeoning state could coexist.[9]

This new view of the functions and scope of the federal government was clearly evident in the presidency of Lyndon B. Johnson, Kennedy's successor. In reminiscing about his six years in office, Johnson often discussed space in the frontier terminology of the Kennedy years. President Johnson called the astronauts "brave pioneers" who "blazed new trails across the untraveled wilderness of space." He depicted the astronauts as the "folk heroes of our time," the best men this nation can produce.[10] New space adventures, he added, evoke the excitement and accomplishment of past eras in which Americans proved themselves once again to be "the sons of pioneers who tamed a broad continent and built the mightiest nation in the history of the world."[11]

But more important, Johnson extended the lessons of the space program to a whole host of other issues and social problems. Space became an example of what a large, sufficiently funded, centralized federal government could accomplish. He stated this point succinctly in his memoirs: "Space was the platform from which the social revolution of the 1960s was launched." In other words, the New Frontier of space launched the Great Society. "If we could send a man to the moon," Johnson explained, "we knew we should be able to send a poor boy to school and to provide decent medical care for the aged."[12]

In the years since the Great Society, Americans have learned that there are limits to the federal government's powers. Ronald Reagan and others have campaigned largely upon that notion. Yet even in the post-Reagan America, many Americans still have strong faith in the ability of the federal government to meet any challenge, to solve any problem. And this faith, in the final analysis, may be the greatest legacy of John F. Kennedy's New Frontier in space.

Notes

Introduction

1. See, for example, William Boot, "NASA and the Spellbound Press," *Columbia Journalism Review* 25 (July/August 1986): 27; David E. Sanger, "NASA Would Shift Some Launches to Private Sector," *New York Times* 12 March 1986: 1; and John Noble Wilford, "America's Future in Space After the Challenger," *New York Times Magazine* 16 March 1986: 32.

2. Walter A. McDougall, *The Heavens and the Earth: A Political History of the Space Age* (New York: Basic, 1985), 305.

3. Ralph G. Martin, *A Hero for Our Time: An Intimate Story of the Kennedy Years* (New York: Macmillan, 1983), 340.

4. John Logsdon, *The Decision to Go to the Moon: Project Apollo and the National Interest* (Cambridge: MIT Press, 1970), 12.

5. Logsdon, *The Decision*, 36–37.

6. Logsdon, *The Decision*, 64.

7. McDougall, *The Heavens*, 221.

8. "The Democratic National Convention Acceptance Address," *Vital Speeches* 26 (1 August 1960), 611. For a discussion of Kennedy's New Frontier theme, see Henry Fairlie, *The Kennedy Promise: The Politics of Expectation* (Garden City, N.Y.: Doubleday, 1973), 81–85; Theodore Sorenson, *Kennedy* (New York: Harper, 1965), 167; and Ronald Carpenter, *The Eloquence of Frederick Jackson Turner* (San Marino, Calif.: Huntington Library, 1983), 84–85.

9. Hugh Sidey, *John F. Kennedy, President* (New York: Atheneum, 1964), 118.

10. McDougall, *The Heavens*, 310.

11. "The President's News Conference of April 21, 1961," in *Public Papers of the Presidents of the United States, John F. Kennedy, 1961* (Washington, D.C.: GPO, 1962), 309.

12. Logsdon, *The Decision*, 93.

13. Lewis J. Paper, *The Promise and the Performance: The Leadership of John F. Kennedy* (New York: Crown, 1975), 368.

14. Logsdon, *The Decision*, 108.

15. Logsdon, *The Decision*, 108.

16. The major figures of the space program also saw ways in which expansion and acceleration of the program could serve their own personal interests. McDougall, *The Heavens*, 322–23.

17. McDougall, *The Heavens*, 4.

18. G. Thomas Goodnight, "The Personal, Technical, and Public Spheres of Argument: A Speculative Inquiry into the Art of Public Deliberation," *Journal of the American Forensic Association* 18 (1982): 220; for further discussion of technical argument, see Thomas B. Farrell, "Knowledge, Consensus, and Rhetorical Theory," *Quarterly Journal of Speech* 62 (1976): 1–14; and Thomas B. Farrell and Thomas Goodnight, "Accidental Rhetoric: The Root Metaphors of Three Mile Island," *Communication Monographs* 48 (1981): 271–300.

19. Walter R. Fisher, "Narration as Human Communication Paradigm: The Case of Public Moral Argument," *Communication Monographs* 51 (1984): 1–22.

20. Fisher, "Narration as Human Communication," 4.

21. Fisher, "Narration as Human Communication," 8.

22. "Special Message to Congress on Urgent National Needs," 25 May 1961, *Public Papers of the President*, 662.

23. "The President's News Conference of October 11, 1961," in *Public Papers of the President, 1961*, 662.

24. McDougall, *The Heavens*, 347.

25. "Colonel Wonderful," *Time* 19 March 1962: 22.

26. "Colonel Wonderful," 22.

27. See, for example, Fairlie, *The Kennedy Promise*, 12–13; Martin, *A Hero*, 339–42; Paper, *The Promise*, 368–69; Arthur M. Schlesinger, Jr., *A Thousand Days: John F. Kennedy in the White House* (Boston: Houghton Mifflin, 1965), 343; Sidey, *John F. Kennedy*, 110–23; and Sorensen, *Kennedy*, 523–29.

28. Stephen Depoe, "Space and the 1960 Presidential Campaign: Kennedy, Nixon, and 'Public Time,'" *Western Journal of Speech Communication* 55 (Spring 1991): 215–33. Theodore Windt wrote two articles on Kennedy's presidential rhetoric. Both, however, focus on Kennedy's treatment of international issues. See "The Presidency and Speeches on International Crises: Repeating the Rhetorical Past" and "Seeking Detente with Superpowers: John F. Kennedy at American University," in *Essays in Presidential Rhetoric*, ed. Theodore Windt (Dubuque: Kendall/Hunt, 1983), 61–69 and 71–84. Robert E. Denton, Jr., and Dan F. Hahn also analyze Kennedy's presidential rhetoric. They examine a single speech on ideology

and technology. *Presidential Communication: Description and Analysis* (New York: Praeger, 1986), 260–70. Roderick P. Hart's study of the verbal style of the presidency includes a few sentences about Kennedy's 13 September 1962 speech at Rice University about space exploration. In the speech, Hart writes, Kennedy did not "gush about the grand adventure of space exploration" (99). Hart says nothing about the possible influence of Kennedy's rhetoric on the space program. Hart's work examines presidential verbal style. *Verbal Style and the Presidency: A Computer-Based Analysis* (Orlando: Academic Press, 1984), 94–126.

29. Janice Hocker Rushing, "Mythic Evolution of 'The New Frontier' in Mass Mediated Rhetoric," *Critical Studies in Mass Communication* 3 (1986): 265–96; Janice Hocker Rushing, " 'The New Frontier' in *Alien* and *Aliens:* Patriarchal Co-Optation of the Feminine Archetype," *Quarterly Journal of Speech* 75 (1989): 1–24; and Janice Hocker Rushing, "Frontierism and the Materialization of the Psyche: The Rhetoric of *Innerspace*," *Southern Communication Journal* 56 (1991): 243–56.

30. Loyd S. Swenson, Jr., James M. Grimwood, and Charles C. Alexander, *This New Ocean: A History of Project Mercury* (Washington, D.C.: NASA, 1966).

31. Vernon Van Dyke, *Pride and Power: The Rationale of the Space Program* (Urbana: University of Illinois Press, 1964).

32. Robert Cirino, "To the Moon: 'There Really Isn't Any Argument,' " in *Don't Blame the People* (Los Angeles: Diversity Press, 1971), 252–63.

33. Edwin Diamond, *The Rise and Fall of the Space Age* (Garden City, N.Y.: Doubleday, 1964), 84–95.

34. Tom Wolfe, *The Right Stuff* (New York: Bantam, 1980).

35. Frank Van Riper, *Glenn: The Astronaut Who Would be President* (New York: Empire, 1983).

36. Robert E. Ostman and William A. Babcock, "Three Major Newspapers' Content and President Kennedy's Press Conference Statements Regarding Space Exploration and Technology," *Presidential Studies Quarterly* 13 (1983): 111–20.

37. Theodore O. Windt, Jr., "Presidential Rhetoric: Definition of a Field of Study," *Central States Speech Journal* 35 (1984): 29.

38. Richard E. Neustadt, *Presidential Power: The Politics of Leadership* (New York: John Wiley, 1960), 10.

39. Windt, "Presidential Rhetoric," 24.

40. James W. Ceasar, Glen E. Thurow, Jeffrey Tulis, and Joseph M. Bassette, "The Rise of the Rhetorical Presidency," *Presidential Studies Quarterly* 10 (1980): 168.

41. See, for example, Ostman and Babcock, and J. Michael Hogan, *The Panama Canal in American Politics: Domestic Advocacy and the Evolution of Policy* (Carbondale: Southern Illinois University Press, 1986), 10.

42. Murray Edelman, *Political Language: Words That Succeed and Policies That Fail* (New York: Academic Press, 1977), 3.

43. Murray Edelman, *Politics as Symbolic Action: Mass Arousal and Quiescence* (Chicago: Markham, 1971), 66.

44. Edelman, *Political Language*, 3.

45. David Zarefsky, *President Johnson's War on Poverty: Rhetoric and History* (University: University of Alabama Press, 1986), 1.

46. Zarefsky, *President Johnson*, 11.

47. Cirino, "To the Moon," 255.

1. The Kennedy Administration's Lunar Campaign

1. U.S. Congress, Senate, Committee on Appropriations, *Independent Offices Appropriations, 1962*, 87th Cong., 1st sess., 1961, 653 (hereinafter Senate, *Independent Offices Appropriations, 1962*). Dryden's comment in 1958 demonstrates that he questioned the value of Project Mercury. Dryden insisted that "tossing a man up into the air and letting him come back" had about "the same technical value as the circus stunt of shooting a young lady from the gun." U.S. Congress, House, Select Committee on Astronautics and Space Sciences, *Astronautics and Space Exploration: Hearings on H.R. 11881*, 85th Cong., 2d sess., 1958, 117.

2. "Special Message to the Congress on Urgent National Needs," 25 May 1961, in *Public Papers of the Presidents, 1961*, 405.

3. Van Dyke, *Pride and Power*, 134.

4. See, for example, "The Interpreters and the Golden Throats," *Newsweek* 8 October 1962: 102, and "Term 'A-OK' Dropped, Astronauts Favor 'Go,'" *New York Times* 21 February 1962: 22.

5. U.S. Congress, House, Committee on Science and Astronautics, *1964 NASA Authorization, Hearings on H.R. 5466*, 88th Cong., 1st sess., 1963, pt. 1, 4 (hereinafter House, *1964 NASA Authorization*).

6. NASA Memo from Brian Duff, Reports and Special Communications Division, to All Public Information Officers, 20 June 1963, 1, National Aeronautics and Space Administration History Office, Washington, D.C. (hereinafter NASA HO).

7. Julian Sheer, "NASA Speakers," 25 June 1963, NASA HO.

8. See, for example, "Emphasis on Space," *Newsweek* 24 September 1962: 18; Marjorie Hunter, "President, Touring Canaveral, Sees a Polaris Fired," *New York Times* 17 November 1963: 44; and John W. Finney, "Senators Uphold Space Fund Cuts," *New York Times* 14 November 1963: 21.

9. John W. Finney, "NASA Considers New Space School," *New York Times* 2 December 1962: 62.

10. See Michael A. Feighan, *Cong. Rec. Appendix* 14 June 1962: A4470.

11. U.S. Congress, House, Committee on Science and Astronautics, *1963 NASA Authorization, Hearings on H.R. 11737*, 87th Cong., 2d sess., 1962, pt. 1, 12 (hereinafter House, *1963 NASA Authorization*).

12. One magazine blasted the timing of the flight. "It may be we are witnessing a phenomenon destined to play an ever larger part in world politics—the use of accurately timed technological spectaculars to extract vast sums from national treasuries and in effect by-pass legislative bodies." "Coincidence?" *Nation* 25 May 1963: 433. Also see, for example, "Under

Whose Moon?" *Time* 31 May 1963: 15; "Astronauts Push for a 7th Flight," *New York Times* 22 May 1963: 1; and John W. Finney, "Space Budget Still Facing Cuts," *New York Times* 26 May 1963: D6.

13. James E. Webb, "Memorandum for the President," 20 November 1962, 9, NASA HO.

14. See, for example, "U.S. Acts to Stem Space Publicity," *New York Times* 2 May 1961: 22; "Like It Comes Up a Seven," *Missiles and Rockets* 15 June 1961: 54; Swenson, Grimwood, and Alexander, *This New Ocean*, 420.

15. See, for example, "Sweeping Inquiry on U.S. Research Pushed in House," *New York Times* 28 July 1963: 49; "Chicago U. Grant by NASA Defended," *New York Times* 28 July 1963: 49; and Stuart H. Loory, "Are We Wasting Billions in Space?" *Saturday Evening Post* 14 September 1963: 15.

16. John D. Morris, "House to Review Charges of Flaws in Space Devices," *New York Times* 5 October 1963: 1, and John W. Finney, "NASA Tempers Criticism," *New York Times* 5 October 1963: 9.

17. Senate, *Independent Offices Appropriations, 1962*, 662.

18. John F. Kennedy, "Memorandum for Vice President," 20 April 1961, NASA HO.

19. "The President's News Conference of April 21, 1961," in *Public Papers of the Presidents, 1961*, 309.

20. "Address at Rice University in Houston on the Nation's Space Effort," 12 September 1962, in *Public Papers of the Presidents of the United States, John F. Kennedy, 1962* (Washington, D.C.: GPO, 1963), 669. Also see "The President's News Conference of August 10, 1961," in *Public Papers of the Presidents, 1961*, 560.

21. Lyndon B. Johnson, "Memorandum for the President: Evaluation of the Space Program," 28 April 1961, 2, NASA HO.

22. James Webb, "Administrator's Presentation to the President," 21 March 1961, 1, NASA HO.

23. Webb, "Administrator's Presentation," 5.

24. James Webb, "Remarks to the American Association of School Administrators," 18 February 1962, 11, NASA HO.

25. House, *1964 NASA Authorization*, pt. 2(a), 1067. Also see House, *1964 NASA Authorization*, pt. 1, 1.

26. U.S. Congress, House, Committee on Science and Astronautics, *1962 NASA Authorization, Hearings on H.R. 6874*, 87th Cong., 1st sess., 1961, pt. 3, 1052 (hereinafter House, *1962 NASA Authorization*). Also see, for example, James Webb, "National Goals in the Space Age," National Aeronautics and Space Administration, *Conference on Space Age Planning* (Washington, D.C.: GPO, 1963), 4, and Wernher von Braun, "The Need to Explore Space," *Cong. Rec.* 26 November 1963: 22837.

27. "Address at Rice University," 670. Economist George Gilder has written that "the space program failed to invent any new technology; its achievements were almost entirely based on the state of the art ten years before." *Wealth and Poverty* (New York: Basic, 1981), 82.

28. "Remarks at the Presentation of NASA's Distinguished Service Medal to Astronaut Alan B. Shepard," 8 May 1961, in *Public Papers of the Presidents, 1961*, 366, and "Remarks at the Presentation of NASA's Dis-

tinguished Service Medal to Dr. Robert R. Gilruth and Col. John H. Glenn, Jr.," 23 February 1962, in *Public Papers of the Presidents, 1962*, 159.

29. U.S. Congress, Senate, Committee on Aeronautical and Space Sciences, *Orbital Flight of John H. Glenn, Jr.*, 87th Cong., 1st sess., 1962, 16 (hereinafter Senate, *Orbital Flight*).

30. Memo for James Webb from Bill Lloyd, "Anticipated Questions from News Media," 21 May 1961, 3, NASA HO.

31. U.S. Congress, House, Committee on Science and Astronautics, *Discussion of Soviet Man-in-Space Shot*, 87th Cong., 1st sess., 1961, 3, and NASA News Release, 24 March 1961, 4, NASA HO.

32. Senate, *Independent Offices Appropriations, 1962*, 651.

33. U.S. Congress, Senate, Committee on Appropriations, *Independent Offices Appropriations, 1964*, 88th Cong., 1st sess., 1963, 1522 (hereinafter Senate, *Independent Offices Appropriations, 1964*).

34. Senate, *Independent Offices Appropriations, 1964*, 1593. For additional comments on science and technology, see, for example, James Webb, "The Role of Government in Scientific Exploration," 14; Homer E. Newell, "Space Science—Earth, Sun, and Stars," 35–47; Edgar M. Cortwright, "Space Science—Moon and Planets," 49–58, in National Aeronautics and Space Administration, *Proceedings of the Second National Conference on the Peaceful Uses of Space* (Washington, D.C.: GPO, 1962); John E. Nagle, "Results of Scientific Research in Space," in *Conference on Space-Age Planning*, 19–26; and Hugh Dryden, National Aeronautics and Space Administration, *Impact of Progress in Space on Science* (Washington, D.C.: GPO, 1962), 4–7.

35. "Address at Rice University," 670.

36. Lyndon B. Johnson, "The New World of Space," in *Proceedings of the Second National Conference*, 30.

37. Webb, "Remarks to American School Administrators," 12.

38. "The President's News Conference of April 24, 1963," in *Public Papers of the President of the United States, John F. Kennedy, 1963* (Washington, D.C.: GPO, 1964), 350. For additional comments on education, see Webb, "The Role of Government in Scientific Exploration," in *Proceedings of the Second National Conference*, 15–16 and 18; Dryden, *Impact of Progress in Space*, 14; Hugh Dryden, National Aeronautics and Space Administration, *Space, the New Frontier* (Washington, D.C.: GPO, 1962), 49; General Curtis LeMay, "Man's Future in Space," *Cong. Rec. Appendix*, 6 March 1963: A1202; and William H. Pickering, "University Research Activities and Their Impact on the National Space Program," in National Aeronautics and Space Administration, *Proceedings of the First National Conference on the Peaceful Uses of Space* (Washington, D.C.: GPO, 1961), 125–30.

39. Lyndon B. Johnson, "Remarks at Space Center Dedication," 14 November 1963, 2, NASA HO.

40. Bernard Schriever, "Space and National Security," *Cong. Rec. Appendix*, 10 October 1961: A8073.

41. House, *1964 NASA Authorization*, pt. 2 (a), 661.

42. Schriever, "Space and National Security," A8074.

43. "Address in Los Angeles at a Dinner of the Democratic Party of California," 18 November 1961, in *Public Papers of the Presidents, 1961,* 734.

44. "The President's News Conference of June 14, 1962," in *Public Papers of the Presidents, 1962,* 485. Also see "The President's News Conference of February 15, 1961," in *Public Papers of the Presidents, 1961,* 94.

45. Lyndon Johnson, "Remarks at a Luncheon in Honor of Gordon Cooper," *Cong. Rec.* 23 May 1963: 9383.

46. House, *1963 NASA Authorization,* pt. 2, 656 and 672.

47. Senate, *Independent Offices Appropriations, 1964,* 1505.

48. House, *1964 NASA Authorization,* pt. 2 (a), 888.

49. Senate, *Independent Offices Appropriations, 1964,* 1516. For additional comments on defense, see, for example, Webb, "The Role of Government in Scientific Exploration," 15; Webb, "National Goals in the Space Age," 1–2; Dryden, *Space, the New Frontier,* 49; Dryden, *Impact of Progress in Space,* 11; von Braun, "The Need to Explore," 22837; and Edward C. Welsh, "Space and National Security," *Cong. Rec. Appendix,* 10 October 1961: A8060.

50. House, *1964 NASA Authorization,* pt. 2 (a), 631.

51. Joseph F. Shea, "Manned Space Flight," U.S., National Aeronautics and Space Administration, *Conference on Space-Age Planning,* 45.

52. House, *1964 NASA Authorization,* pt. 2 (a), 132.

53. James Webb, "NASA Distinguished Service Awards," NASA News Release, 27 May 1962, 2, NASA HO. Also see Louis B. C. Fong, "The NASA Program of Industrial Applications," in *Proceedings of the First National Conference,* 185.

54. Memo to James Webb from Bill Lloyd, 4.

55. John F. Kennedy, "Opening Remarks," in *Proceedings of the First National Conference,* vii.

56. Senate, *Independent Offices Appropriations, 1962,* 649.

57. Lyndon B. Johnson, "American Institute of Aeronautics and Astronautics Second Manned Space Flight Meeting," 23 April 1963, 4, NASA HO.

58. Johnson, "The New World of Space," 31.

59. Senate, *Independent Offices Appropriations, 1964,* 1566.

60. Lyndon B. Johnson, "Speech at the National Rocket Club Dinner," 22 March 1963, 5 and 2, NASA HO.

61. House, *1962 NASA Authorization,* pt. 3, 1053.

62. Johnson, "National Rocket Club Dinner," 2.

63. Memo to James Webb from Bill Lloyd, 4.

64. Senate, *Independent Offices Appropriations, 1962,* 649.

65. Senate, *Independent Offices Appropriations, 1964,* 1516.

66. U.S. Congress, House, Committee on Appropriations, *Independent Offices Appropriations for 1963,* 87th Cong., 2d sess., 1962, pt. 3, 425 (hereinafter House, *Independent Offices Appropriations, 1963*).

67. Johnson, "The New World of Space," 31.

68. Lyndon B. Johnson, "Remarks at the Goddard Memorial Award Dinner," 19 March 1962, 5, NASA HO. For other comments on the economics, see, for example, Webb, "The Role of Government in Scientific Explora-

tion," 19; James Webb, National Aeronautics and Space Administration, *Space, the New Frontier* (Washington, D.C.: GPO, 1963), 2; Webb, "National Goals in the Space Age," 3; James Webb, "Speech to the Upper Midwest Research and Development Council," in *Cong. Rec.* 23 February 1963: 2932. Dryden, *Space, the New Frontier*, 49; Dryden, *Impact of Progress in Space*, 11–12; von Braun, "The Need to Explore," 22837; Edward C. Welsh, "Space and the National Economy," *Cong. Rec.* 9 July 1963: A4240; and Wernher von Braun, Letter to Representative George Andrews, *Cong. Rec.* 10 October 1963: 19251.

69. "Address at Rice University," 670.

70. Letter to Representative George Andrews in *Cong. Rec.* 10 October 1963: 19250–51.

71. Webb, "Remarks to American School Administrators," 10.

72. Johnson, "The New World of Space," 30.

73. Kennedy, "Opening Remarks," vii.

74. James Webb, in *Proceedings of the First National Conference*, 99. Also see, for example, Lyndon Johnson, "National Rocket Club Dinner," 3.

75. Johnson, "Goddard Memorial," 4.

76. Johnson, "American Institute of Aeronautics," 4. Also see Webb, "American School Administrators," 7.

77. House, *1962 NASA Authorization*, pt. 3, 1044. For additional comments, see, for example, Webb, "The Role of Government in Scientific Exploration," 20–21; Webb, *Space, the New Frontier*, 1963, 5; Webb, "National Goals in the Space Age," 3; von Braun, "The Need to Explore," 22837; Lyndon B. Johnson, "Speech at the Board of City Development Dinner," in *Cong. Rec.* 6 March 1963: 3593; and Lyndon B. Johnson, "National Rocket Club Dinner," 3.

78. House, *Independent Offices Appropriations, 1963*, 543.

79. House, *1962 NASA Authorization*, 791.

80. U.S. Congress, House, Committee on Appropriations, *Independent Offices Appropriations, 1962*, 87th Cong., 1st sess., 1961, pt. 2, 1092–93, and Senate, *Orbital Flight*, 24.

81. "Remarks upon Presenting the NASA Distinguished Service Medal to Astronaut L. Gordon Cooper," 21 May 1963, in *Public Papers of the Presidents, 1963*, 417.

82. House, *1962 NASA Authorization*, pt. 1, 77.

83. Senate, *Independent Offices Appropriations, 1964*, 1521.

84. Senate, *Independent Offices Appropriations, 1964*, 1620.

2. The Kennedy Administration and the New Frontier

1. Henry Nash Smith, *Virgin Land: The American West as Symbol and Myth* (Cambridge: Harvard University Press, 1978), 17.

2. McDougall, *The Heavens*, 305.

3. Janice Hocker Rushing, "Mythic Evolution," 265. For other studies on the frontier, see Janice Hocker Rushing, "The Rhetoric of the Western

Myth," *Communication Monographs* 50 (1983): 14–32; Sarah Hankins, "Archetypal Alloy: Reagan's Rhetorical Image," *Central States Speech Journal* 33 (1983): 33–43; Ronald Carpenter, "Frederick Jackson Turner and the Rhetorical Impact of the Frontier Thesis," *Quarterly Journal of Speech* 63 (1977): 117–29; and Carpenter, *The Eloquence.* For information on narrativity, see, for example, Walter Fisher, "Narration as Human Communication," 1–22; Alasdair MacIntyre, *After Virtue: A Study in Moral Theory* (Notre Dame, Ind.: University of Notre Dame Press, 1981); and Hayden White, "The Value of Narrativity in the Representation of Reality," *Critical Inquiry* 7 (1980): 5–27.

4. Richard Slotkin, *The Fatal Environment: The Myth of the Frontier in the Age of Industrialization, 1800–1890* (New York: Atheneum, 1985), 34.

5. Rushing, "Mythic Evolution," 71.

6. Rushing, "Mythic Evolution," 72.

7. Martin, *A Hero,* 340; "Special Message to the Congress on Urgent National Needs," 25 May 1961, in *Public Papers of the Presidents, 1961,* 405.

8. John F. Kennedy, "Opening Remarks," in National Aeronautics and Space Administration, *Proceedings of the First National Conference,* vii.

9. Zarefsky, *President Johnson,* 17.

10. Rushing, "Mythic Evolution," 78–86.

11. Wolfe, *The Right Stuff.*

12. "Remarks in St. Louis to Employees of the McDonnell Aircraft Corporation," 12 September 1962, in *Public Papers of the Presidents, 1962,* 672.

13. "Address at Rice University," 671.

14. "Remarks upon Presenting the NASA Distinguished Service Medal to Astronaut L. Gordon Cooper," 21 May 1963, in *Public Papers of the Presidents of the United States: John F. Kennedy, 1963* (Washington, D.C.: GPO, 1964), 417.

15. House, *1964 NASA Authorization,* and John Johnson, "The New Frontier of Space," Address to the New York Patent Law Association, 24 March 1961, 2, NASA HO.

16. Johnson, "New Frontier of Space," 2.

17. "Presentation of NASA Distinguished Service Awards," 24 May 1962, 4, NASA HO.

18. House, *1962 NASA Authorization.*

19. Vice President Lyndon B. Johnson, "Goddard Memorial Awards Dinner," 16 March 1962, 6, NASA HO.

20. "Address at Rice University," 670.

21. "NASA Distinguished Service Medal to Gordon Cooper," 198.

22. "Remarks Intended for Delivery to the Texas Democratic State Committee in the Municipal Auditorium in Austin," 22 November 1963, in *Public Papers of the Presidents, 1963,* 897, and "Remarks in San Antonio at the Dedication of the Aerospace Medical Health Center," 21 November 1963, in *Public Papers of the President, 1963,* 882.

23. "Vice President Lyndon B. Johnson, Space Center Dedication," 14 November 1963, 1, NASA HO.

24. "Remarks of Vice President Johnson, American Institute of Aeronautics and Astronautics Second Manned Space Flight Meeting," 24 April

1963, 5, NASA HO. Also see, for example, James Webb, "The Role of Government in Scientific Exploration," 12; James Webb, "Fifth Anniversary of Tracking Goddard Space Flight Center," *Cong. Rec.* 31 January 1963: 2382; Roger Revelle, "Sailing in New and Old Oceans," in *Proceedings of the Second National Conference,* 25; Wernher von Braun, "Launch Vehicles," in *Proceedings of the First National Conference,* 72; Webb, *Space, the New Frontier,* 4; National Aeronautics and Space Administration, *Conference on Space Age Planning,* ix; Wernher von Braun, "The Need to Explore," *Cong. Rec.* 26 November 1963: 22835; Curtis LeMay, "Man's Future in Space," A1203; and Lyndon Johnson, "Board of City Development Dinner," *Cong. Rec.* 22 February 1963: 3593.

25. Scott Carpenter, in *We Seven* (New York: Simon and Schuster, 1962), 368; Alan Shepard, in *We Seven,* 68. See also Deke Slayton, 71.

26. John Glenn, "We're Going Places No One Has Ever Gone in a Craft No One Has Ever Flown," *Life* 27 January 1961: 46.

27. John Glenn, in *We Seven,* 26–27.

28. Glenn, in *We Seven,* 27.

29. Virgil Grissom, in *We Seven,* 59.

30. Rushing, "Mythic Evolution," 283.

31. *Cong. Rec.* 10 October 1963: 19250.

32. House, *Discussion of Soviet Man-in-Space Shot,* 22.

33. House, *1964 NASA Authorization,* pt. 3, 91 and 180. Also see, for example, Robert Gilruth, "Mercury and Apollo: Results and Plans," in *Proceedings of the First National Conference,* 110; James Webb, "Luncheon Honoring Delegates to the Youth Science Camp," *Cong. Rec.* 19 July 1963: A5057; James Webb, "Upper Midwest Research and Development Dinner," *Cong. Rec.* 29 January 1963: 2933; and James Webb, in National Aeronautics and Space Administration, *One-Two-Three and the Moon: Projects Mercury, Gemini, and Apollo of America's Manned Space Flight Program* (Washington, D.C.: GPO, 1963), 28.

34. Glenn, in *We Seven,* 24, and Shepard, in *We Seven,* 164.

35. "The President's News Conference of August 10, 1961," in *Public Papers of the Presidents, 1961,* 560.

36. Johnson, "National Rocket Club Dinner," 5.

37. *Cong. Rec.,* 3 April 1962: 5769.

38. James Webb, "Address," in *Proceedings of the First National Conference,* 98.

39. Rushing, "Mythic Evolution," 272.

40. Slayton, in *We Seven,* 72.

41. Carpenter, in *We Seven,* 45.

42. John Glenn, "Space Is at the Frontier of My Profession," *Life* 14 September 1959: 37.

43. Rushing, "Mythic Evolution," 279.

44. Walter Schirra, "There Won't Be Time to Send for the Manual," *Life* 14 September 1959: 37.

45. Grissom, in *We Seven,* 55.

46. Slayton, in *We Seven,* 71.

47. "Astronauts Press Conference," 16 September 1959, 24, NASA HO.

48. "Astronauts Press Conference," 24.

49. Schirra, "There Won't Be Time," 37.

50. Grissom, in *We Seven*, 55.

51. Slayton, in *We Seven*, 71. See also Gordon Cooper, "I've Got the Normal Desire to Go a Little Higher," *Life* 14 September 1959: 28.

52. "Press Conference: Astronaut Program Outlined," 12 May 1959, 15 NASA HO. Also see, for example, Dryden, *Impact of Progress in Space*, 11; James Webb, "National Goals in the Space Age," in *Conference on Space Age Planning*, 5; James E. Elms, "Manned Spacecraft," in *Conference on Space Age Planning*, 47; and David H. Stoddard, "The Human Factor in Manned Space Flight," in *Conference on Space Age Planning*, 67.

53. Swenson, Grimwood, and Alexander, *This New Ocean*, 487.

54. Memo, Walt Bonney to Herb Rosen, "Names of NASA Vehicles," 12 January 1960, NASA HO.

55. "Press Conference: Astronaut Program," 18.

56. News Release, "Mercury Redstone 3," 26 April 1961, 2, NASA HO.

57. News Release, "MR-4 Design Changes," 16 July 1961, 3, NASA HO.

58. Alan Shepard, "The Pilot's Story," *National Geographic* September 1961: 441.

59. Alan Shepard, "The Astronaut's Story of the Thrust into Space," *Life* 19 May 1961: 28.

60. Loudon Wainwright, "The Three Chosen for First Space Ride," *Life* 3 March 1961: 30.

61. Wainwright, "The Three Chosen," 28.

62. "National Aeronautics and Space Administration News Conference," 22 July 1961, 8, NASA HO.

63. Virgil Grissom, "Hero Admits He Was Scared," *Life* 4 August 1961: 98.

64. Virgil Grissom, "It Was a Good Flight and a Great Float," *Life* 28 July 1961: 28.

65. "Titov's Triumph in 17 Orbits," *Life* 18 August 1961: 43.

66. John Glenn, "If You're Shook Up, You Shouldn't Be There," *Life* 9 March 1962: 29.

67. Glenn, in *We Seven*, 339.

68. Robert Voas, "John Glenn's Three Orbits in Friendship 7," *National Geographic* June 1962: 793.

69. Glenn, in *We Seven*, 252–53.

70. Swenson, Grimwood, and Alexander, *This New Ocean*, 407.

71. John Dille, in *We Seven*, 11.

72. Dille, in *We Seven*, 11.

73. Senate, *Orbital Flight*, 13; see also Glenn, "I'll Have to Hit That Keyhole in the Sky," *Life* 8 December 1961: 50.

74. "Remarks at the Annual Presidential Prayer Breakfast," 1 March 1962, in *Public Papers of the Presidents, 1962*, 175–76.

75. Glenn, "If You're Shook Up," 26–27.

76. Senate, *Orbital Flight*, Appendix H, "Transcript of Colonel Glenn's Press Conference," 96.

77. Senate, *Orbital Flight*, Appendix H, 91.

78. *Cong. Rec.*, 26 February 1962: 2902.
79. Glenn, in *We Seven*, 41; see also "Space Is at the Frontier," 38.
80. *Cong. Rec.*, 26 February 1962: 2902.
81. *Cong. Rec.*, 26 February 1962: 2902.
82. John Glenn, "I'll Have to Hit That Keyhole," 50.
83. Glenn, in *We Seven*, 24; see also Senate, *Orbital Flight*, 9.
84. Glenn, "If You're Shook Up," 29.
85. Glenn, in *We Seven*, 351.
86. House, *1963 NASA Authorization*, pt. 1, 146.
87. Senate, *Orbital Flight*, Appendix G, "MA-6 Press Conference," 78.
88. House, *1963 NASA Authorization*, pt. 2, 459.
89. House, *1963 NASA Authorization*, pt. 1, 33.
90. House, *1963 NASA Authorization*, pt. 1, 33.
91. Senate, *Orbital Flight*, 25.
92. House, *Independent Offices Appropriations, 1963*, pt. 3, 713.
93. Dille, in *We Seven*, 12.
94. McDougall, *The Heavens*, 347.
95. "Remarks at the Presentation of NASA's Distinguished Service Medal to Dr. Robert R. Gilruth and Col. John H. Glenn, Jr.," 23 February 1962, *Public Papers of the Presidents, 1962*, 159.
96. "Presentation of NASA Distinguished Service Awards," and "MA-7 Press Conference," 27 May 1962, 7, NASA HO.
97. Walter Schirra, "A Real Breakthrough—The Capsule Was All Mine," *Life* 26 October 1962: 39.
98. Schirra, "A Real Breakthrough," 39.
99. Edwin Diamond proposed that beginning in 1963, "a curious shift of opinion took place" that eventually "threatened to sweep away the lunar program." In 1961, Diamond observed, America seemed ready for a moon shot. By 1963 many asked "why go at all." *The Rise and Fall*, 35; see also Logsdon, *The Decision*, 168.
100. News Release, "Mercury-Atlas 9," 10 May 1963, 1 and 4, NASA HO. Interestingly, two weeks earlier, George Low testified before the House Committee on Science and Astronautics that the Mercury program had "no guidance equipment." House, *1964 NASA Authorization*, pt. 2 (a), 796.
101. Gordon Cooper, "Everyone Was in a Scare, But I Was Secretly Pleased," *Life* 7 June 1963: 30.
102. It was only the third time that President Kennedy had personally presented the medal. He had previously presented it to Shepard and Glenn. The other astronauts received their medals from NASA's administrator, James Webb.
103. "NASA Distinguished Service Medal to Astronaut L. Gordon Cooper," 416.
104. "Remarks upon Presenting the Collier Trophy to the First U.S. Astronauts," 10 October 1963, *Public Papers of the Presidents, 1963*, 775.
105. "NASA Distinguished Service Medal to L. Gordon Cooper," 417.
106. Elliot See, in "Head over Heels for What's Out There," *Life* 27 September 1963: 86.
107. Neil Armstrong, in "Head Over Heels," 84.
108. Edward White, in "Head Over Heels," 89.

3. Media Coverage of the Space Program: A Reflection of Values

1. "The Interpreters," 101.
2. "The Interpreters," 101.
3. "The Interpreters," 102.
4. Cirino, "To the Moon," 255. Walter McDougall indicts the press too, proposing that the Kennedy administration sold the space race with the "aid of sympathetic media." *The Heavens*, 221. Some writers of the time attacked the popular press's coverage of the astronauts, particularly coverage of John Glenn. See Jerry Talmer, "Rockets to the Moon," *Village Voice* 1 March 1962: 4, and "Packaging Bravery," *New Republic* 26 March 1962: 2.
5. Ostman and Babcock, "Three Major Newspapers," 117.
6. Ostman and Babcock, "Three Major Newspapers," 113. Cirino examined the front page of the *New York Times* and *Los Angeles Times*, June and July 1969, and the first three pages of the *Honolulu Star-Bulletin* from 12 January through 31 May 1969. In addition, he examined NBC and CBS network newscast coverage from 10 July through 10 September 1969 and ABC and Mutual network radio news coverage from 22 August through 22 October 1969. "To the Moon," 257–58.
7. Cirino extrapolates from his studies of press coverage in 1969 that press coverage during the previous ten years is similar. He does examine the covers of *Life*, *Time*, and *Newsweek* from 1962 through July 1969 and proposes that only four of the sixty-three covers were unfavorable. Yet he does not examine the content of the magazines. "To the Moon," 258.
8. Ostman and Babcock limited their examination to three newspapers: the *New York Times*, the *Christian Science Monitor*, and the *Wall Street Journal*. Furthermore, they looked at only the issues that appeared one day prior to, the day of, and one day after President John F. Kennedy held a press conference in which he made statements about space technology and exploration and in which he mentioned the Soviet Union. Of the twenty-two press conferences in which Kennedy made comments about both subjects, the authors randomly chose eleven. The authors do not adequately explain their criteria for selection.
9. Ostman and Babcock, "Three Major Newspapers," 113.
10. The authors reported that of the twenty-seven articles in which coders detected bias, twelve were in the anti-Kennedy direction. It is unclear whether the anti-Kennedy bias reflected a rejection of the whole program or a rejection of only the manned program.
11. *Life*'s contract with the astronauts may explain why few articles critical of the program found their way into the pages of the magazine. The popular science magazines, probably because of lack of funds, provided little coverage of the space program. When they did, the coverage was positive. *Popular Science* signed Wernher von Braun to write a column in the magazine in which he answered readers' questions about space. See Robert Crossley, "Our Most Important Announcement in 91 Years," *Popular Science* January 1963: 55, and Wernher von Braun, "Why I Am Writing for Popular Science," *Popular Science* January 1963: 56. See also Joan Steen,

"Why the Moon Is a Must," *Popular Science* September 1963: 92–188, and Alden P. Armagnac, "Ten Toughest Problems of Putting a Man on the Moon," *Popular Science* February 1962: 118–21. Moreover, *Look*, one of the best-selling magazines of the time, carried too few articles to mention.

12. A few articles present the testimony of critics of manned flight, questioning the military, propaganda, economic, and scientific value of sending a man instead of an unmanned satellite. See particularly "Space Surge," *Time* 6 June 1960: 62–63; "Man in Space," *Time* 5 September 1960; and "Should Future Astronauts Be Cerebral?" *Time* 9 November 1962. By mid-1963, *Time* had begun including the arguments of the critics of a manned lunar landing. Some of the articles pointed out that the critics favored unmanned space exploration as a predecessor to manned flight. See "To Moon or Not to Moon," *Time* 31 May 1963: 53; "Still Moonward Bound," *Time* 19 July 1963: 19; "Women Are Different," *Time* 28 July 1963: 26; and "The Grandstands Are Emptying for the Race to the Moon," *Time* 4 October 1963: 64–65.

13. Richard Witkin, "U.S. Space Flight Scheduled Today," *New York Times* 2 May 1961: 20:6; John W. Finney, "What's Next in Space," *New York Times* 14 May 1961: D12; "Astronaut Guides Space Maneuvers," *New York Times* 25 May 1962: 17; John W. Finney, "Pilots Will Control Gemini Spacecraft," *New York Times* 15 October 1962: 1 and 5; Richard Witkin, "Training for Space," *New York Times* 18 June 1963: 3.

14. See, for example, Robert K. Plumb, "Robots Suggested as Explorers of the Moon and Sea Bottoms," *New York Times* 30 September 1960: 29; William Laurence, "Goals in Space," *New York Times* 30 April 1961: D9; "Scientists Polled on Moon Projects," *New York Times* 23 July 1961: 36; "Astronomers Poll Cool to Manned Moon Trip," *New York Times* 1 August 1961: 3; Richard Witkin, "Flight by Schirra Viewed as Proving Pilots' Space Role," *New York Times* 5 October 1962: 1; and "Space Goals Put Strain on Budget," *New York Times* 5 November 1962: 1.

15. "Reaching for the Moon," *New York Times* 29 January 1963: 6; James Reston, "The Man on the Moon and the Men on the Dole," *New York Times* 5 April 1963: 46; "Confusion in Space," *New York Times* 7 April 1963: D12; James Reston, "What Government Official Do You Believe," *New York Times* 24 April 1963: 34; "The Moon Ticket," *New York Times* 19 June 1963: 36; "Controlling the Space Race," 28; "Down to Earth in Space," *New York Times* 29 July 1963: 18; "Bungling in Space," 42; and Arthur Krock, "Racing to the Moon," *New York Times* 20 October 1963: D11.

16. See, for example, "Space Surge," *Newsweek* 6 June 1960: 62–63; "Our Rockets Are Flying, Astronauts Poised," *Newsweek* 2 January 1961: 42; "Moon Madness," *Newsweek* 6 May 1963: 81; "Men and the Moon," *Newsweek* 29 July 1963: 63; and "De-Accelerating," *Newsweek* 8 July 1963: 62. Also see Harland Manchester, "The Senseless Race to Put Man in Space," *Reader's Digest* May 1961: 67.

17. Herbert J. Gans, "The Messages Behind the News," *Columbia Journalism Review* 17 (January/February 1979), 40.

18. Gans, "The Messages," 43.

19. "The Grandstands Are Emptying for the Race to the Moon," *Time* 4 October 1963: 64.

20. Stuart H. Loory, "Are We Wasting Billions in Space?" *Saturday Evening Post* 14 September 1963: 13.

21. "The Grandstands Are Emptying," 64.

22. John W. Finney, "Space Program: Too Expensive?" *New York Times* 12 May 1963: D8. Also see "Space Goals Put Strain on Budget," 6, and John W. Finney, "Space Spending Called Austere," *New York Times* 26 February 1963: 5.

23. Loory, "Are We Wasting Billions?" 14.

24. "Controlling the Space Race," 28.

25. John W. Finney, "House Unit Votes Space Fund Cuts," *New York Times* 26 June 1963: 12. See also "Down to Earth in Space," 18.

26. Gans, "The Messages," 41.

27. Gans, "The Messages," 41.

28. "Bungling in Space," 42. Also see notes 29–33 on governmental waste.

29. "The Race Is Over," *Newsweek* 4 November 1963. See also Loory, "Are We Wasting Billions?" 15; "U.S. Space Program: Streamlined and Stretched Out," *Newsweek* 17 December 1962: 72; and "Down to Earth in Space," 18.

30. "Ten Per Cent of the Moon, Too?" *New York Times* 28 August 1963: 32; "Commercialism to the Moon," *New York Times* 20 February 1963: 10; "Commercialism in Space," *New York Times* 1 June 1963: 20; "Let Us Explore the Stars," *New York Times* 19 September 1963: 26; "The Insensitives," *New York Times* 25 October 1963: 30; "America's New Astronauts," *New York Times* 18 September 1962: 38; and James Reston, "Even Up in Outer Space, Money Is a Problem," *New York Times* 22 August 1962: 32. Ironically, in 1971, the newspaper bought syndication rights for articles by *Apollo 15* astronauts David R. Scott, James B. Irwin, Alfred M. Worden, and Harrison Schmitt. See Col. David R. Scott, "Finding the Golden Egg," *New York Times* 13 August 1971: 1, and "Irwin and Worden: Two Views of the Moon," *New York Times* 14 August 1971: 1. Also see Robert Sherrod, "The Selling of the Astronauts," *Columbia Journalism Review* 12 (May/June 1973): 16–25.

31. "Publisher Drops 3-Million Offer to Astronauts for their Stories," *New York Times* 9 July 1963: 13. See also John W. Finney, "U.S. Will Allow All Astronauts to Sell Stories of Space Trip," *New York Times* 16 September 1962: 40; John W. Finney, "Kennedy Backs Sale of Space Memoirs," *New York Times* 24 November 1962: 8; John W. Finney, "Astronauts' Stories Sought for $800,000," *New York Times* 26 August 1963: 2; "Astronauts' Sale of Stories Scored," *New York Times* 2 July 1963: 10; John W. Finney, "$3 Million Offered to Sixteen Astronauts," *New York Times* 9 February 1963: 4; John W. Finney, "NASA Reviewing Astronauts' Deal," *New York Times* 31 May 1963: 3; and "Seven Astronauts to Share Rewards of Space Trip," *New York Times* 20 July 1959: 25.

32. "Astronauts' Ordeal: Greater on the Ground," *Newsweek* 8 October 1962: 31. See also "Spaceman's Ordeal," *Newsweek* 5 February 1962, 18; "The Long Wait," *Newsweek* 2 September 1963: 54; "If You're an Astronaut," *Newsweek* 7 September 1959: 60; "Orbital Ills," *Newsweek* 18 February 1963: 58; and "New Boys and Old," *Newsweek* 4 March 1963: 51.

The closest *Time* came to criticizing the contract was a summary of the positions of two other publications, one for the contract and one against. "Scrubbed on the Pad," *Time* 19 July 1963: 40. See also "Fringe Benefits from Space," *Time* 27 September 1963: 40.

33. "Spaceman's Ordeal," *Newsweek* 5 February 1962: 18, and "The Long Wait," *Newsweek* 2 September 1963: 54.

34. "Project Mercury: Late, Late Show," *Newsweek* 12 February 1962: 55.

35. "Bungling in Space," 42.

36. "Changing Vistas—East, West, and South," *Newsweek* 8 October 1962: 28.

37. "Frictions in Space," *New York Times* 3 February 1962: 20. Also see Edwin Diamond, "That Moon Trip: Debate Sharpens," *New York Times Magazine* 28 July 1963: 10; "Changing Vistas," 28; Loory, "Are We Wasting Billions?" 15; "Still Moonward Bound," 19; John W. Finney, "Funds for Space in Peril in House," *New York Times* 25 September 1963: 15; "Bungling in Space," *New York Times* 8 October 1963: 42; and "Frictions in Space," *New York Times* 3 February 1962: 20.

38. "Still Moonward Bound," *Time* 19 July 1963: 19.

39. "Changing Vistas," 28.

40. "No Pork Barrel in Space," *New York Times* 25 August 1963: D12; "Controlling the Space Race," *New York Times* 28 June 1963: 28; "The Senator Delivers," *New York Times* 3 August 1963: 16; "Bungling in Space," 42; John Finney, "Boston Space Lab Snagged in House," *New York Times* 14 June 1963: 12. See also note 41. The values clustered around altruistic democracy influenced reporting of other events, like a proposed $400,000 NASA contract with the Columbia University Graduate School of Journalism to improve public understanding of science and how space information was disseminated. The university called off the grant when critics began speculating that NASA's true motive was to improve its public relations. See, for example, "Chicago U. Grant," 49; "Sweeping Inquiry," 49; and Loory, "Are We Wasting Billions?" 15.

41. "The Senator Delivers," *New York Times* 3 August 1963: 16.

42. W. Lance Bennett, *News: The Politics of Illusion*, 2d ed. (New York: Longman, 1988), 26.

43. Gans, "The Messages," 43.

44. "John Glenn: One Machine That Worked Without Flaw," *Newsweek* 5 March 1962: 24.

45. "On the Bazoo," *Newsweek* 27 May 1963: 61. The magazine described the flights of Alan Shepard, Wally Schirra, and Gordon Cooper in the same way, emphasizing the "fate of the single man." See "Everything A-Okay," *Newsweek* 15 May 1961: 28; "The Promise of Sigma 7," *New York Times* 10 October 1962: 38; and "Laurels," *New York Times* 22 May 1963: 40.

46. See "Rendezvous with Destiny," *Time* 20 April 1959: 17, and "The Seven Chosen," *Time* 20 April 1959: 18. In the same issue, Norman Barr describes the astronaut as "the one single sample from all the billions of men that populate the earth." "A New Human Experience," *Time* 20 April 1959: 20.

47. "Spaceman's Ordeal," 15, and Raymond Moley, "What's Back of a Hero," *Newsweek* 9 April 1962: 116. See also "Carpenter Committed," *Newsweek* 28 May 1962: 21.

48. Richard Witkin, "An Orbital Flight of Up to Six Days Being Considered," *New York Times* 18 May 1963: 8.

49. "Taciturn Astronaut," *New York Times* 22 July 1961: 8. For similar descriptions see "Freedom's Flight," *Time* 12 May 1961: 54; "The New Ocean," *Time* 2 March 1962: 11; "Spaceman's Ordeal," 18; Raymond Moley, "What's Back of a Hero," 116; "New York Pauses to 'Watch' Glenn," *New York Times* 21 February 1962: 1; and John W. Finney, "The Race in Space Still Has a Long Way to Go," *New York Times* 7 May 1961: D8.

50. Richard Witkin, "Project Apollo: Man's Race for Moon," *New York Times* 30 July 1962: 1. Also see, for example, John W. Finney, "Astronauts Can't Be Automated," *New York Times Magazine* 5 April 1963: 117; "Rendezvous with Destiny," 7. The press often compared Glenn to Lindbergh. See, for example, "Space: The New Ocean," *Time* 2 March 1962: 18; "Colonel Wonderful," 22; and Raymond Moley, "What's Back of a Hero," 116.

51. "Men in Space," *New York Times* 11 April 1959: 20. The article also describes the astronaut as the "Columbus of space."

52. "Rendezvous with Destiny," 7.

53. James Reston, "When Carpenter Got Back He Asked for Water," *New York Times* 27 May 1962: D12. See also James Reston, "The Sky's No Longer the Limit," *New York Times* 12 April 1959: D2; "Scott Carpenter's Ride," *New York Times* 25 May 1962: 32; John W. Finney, "The Race in Space," D8; John W. Finney, "Astronauts Give View," *New York Times* 1 March 1962: 1; Arthur C. Clarke, "Down-to-Earth Survey of Space," *New York Times Magazine* 5 November 1961: 40; "A Man for the Job," *New York Times* 29 May 1962: 30; Bernard Lovell, "The Greatest Challenge to Man," *New York Times Magazine* 21 April 1963: 48; and John W. Finney, "Astronauts Can't," 117.

54. Reporting on the astronauts' initial press conference, Reston remarked that the astronauts "talked of the heavens the way the old explorers talked of the unknown seas." He also compared the astronauts to "Walt Whitman's pioneers." See James Reston, "The Sky's No Longer," D2.

55. "Men in Space," 20.

56. "How Seven Were Chosen," *Newsweek* 20 April 1959: 64.

57. Barr, "A New Human Experience," 19.

58. Even the title of the article suggested an interpretation. "Rendezvous with Destiny," 7.

59. Kenneth Crawford, "The Politics of Space," *Newsweek* 8 October 1962: 31.

60. John W. Finney, "Astronauts Can't," 117.

61. For descriptions of space exploration as an adventure story, see "The Effort in Space," *New York Times* 14 September 1962: 30; "The Coming Countdown for the Astronaut," *Newsweek* 12 December 1960: 57; Lovell, "The Greatest Challenge," 48; "The Promise," 38; "Men on the Moon," *Newsweek* 19 July 1961: 62; Armagnac, "Ten Toughest Problems," 119;

Francis V. Drake, "We're Running the Wrong Race with Russia!" *Reader's Digest* August 1963: 50; and Dan Q. Posin, "Race to the Moon," *Popular Mechanics* August 1959: 64. For descriptions of the moon as the astronauts's target, see, for example, "Hitting the Moon," *Newsweek* 21 September 1959: 80; Posin, "Race to the Moon," 64; "Man and the Moon," *New York Times Magazine* 29 March 1959: 13; "The Promise," 38; and "Sweet Little Bird," *Time* 12 October 1962: 46. The media also described space as unknown and hostile. See, for example, "New Boys and Old," 50; "Saga of the Liberty Bell," *Time* 28 July 1961: 34; Robert Cahn, "Can the Astronaut Come Back Alive?" *Saturday Evening Post* 15 April 1961: 34; Hanson W. Baldwin, "A Space Man's Shadow," *New York Times* 8 August 1961: 12; "Reaching for the Moon," *Time* 10 August 1962: 57; and Witkin, "Project Apollo," 1; Walter Sullivan, "Apollo: Astronauts Will Learn Makeup of Moon," *New York Times* 31 July 1962: 12; and "Men in Space," 20. For descriptions of the Soviets as antagonists, see, for example, Francis V. Drake, "We're Running the Wrong Race," 54, and Posin, "Race to the Moon," 65. See also "Still Moonward Bound," 20, and "The High Ground," *Time* 24 August 1962: 7.

62. Bennett, *News*, 26.

63. "The Astronauts," *New York Times* 1 March 1962: 30. See, for example, "Men in Space," 20; Reston, "The Sky's No Longer," D2; and "Rendezvous near the Moon," *New York Times* 8 July 1962: D8.

64. "New Boys and Old," 50.

65. Dan Q. Posin, "An Eye on Space," *Popular Mechanics* March 1959: 96. For a similar description, see "Rendezvous with Destiny," 7.

66. Armagnac, "Ten Toughest Problems," 119; "Rendezvous with Destiny," 7; Baldwin, "A Space Man's Shadow," 5; "The Journey of Enos," *New York Times* 3 December 1961: D8; and Finney, "The Race in Space," D8.

67. "A Man for the Job," 30. See also "Playing it Cool: The Astronaut and His Wife," *Newsweek* 4 June 1962: 22.

68. "Great Gordo," *Time* 24 May 1963: 21.

69. "Trouble at the Controls," *Newsweek* 3 June 1963: 74. The magazine applies the same description to the new astronauts in "New Boys and Old," 50.

70. "Colonel Wonderful," 22.

71. James Reston, "Is the Moon Really Worth John Glenn?" *New York Times* 25 February 1962: D10.

72. "Colonel Wonderful," 22.

73. "John Glenn: One Machine," 19.

74. Arthur Krock, "Another Talent Revealed Today by Colonel Glenn," *New York Times* 27 February 1962: 32, and James Reston, "Is the Moon Really Worth," D10. See, for example, "The Crew Is Go," *New York Times* 26 January 1962: 1; "Nerveless?" *Time* 23 February 1962: 26; and "John Glenn: One Machine," 20.

75. "Spaceman's Ordeal," 14.

76. "Space: The New Ocean," 14.

77. "John Glenn: One Machine," 20.

78. Richard Witkin, "Astronaut's Flight is Reset," *New York Times* 19 July 1961: 8.

79. "Space: The New Ocean," 18. Moley, "What's Back of a Hero," 116.

80. "Colonel Wonderful," 22. See also, "First American in Orbit," *New York Times* 21 February 1962: 20.

81. Gay Talese, "Glenn Family Calm Amid Cheers and Confetti," *New York Times* 24 February 1962: 13.

82. "Space: The New Ocean," 17.

83. "How Seven Were Chosen," 65.

84. "John Glenn: One Machine," 20.

85. See, for example, Richard Witkin, "Fleet in Position for Orbital Shot," *New York Times* 22 January 1962: 11; Richard Witkin, "Glenn Scheduled to Orbit Earth Today," *New York Times* 27 February 1962: 1; "Glenn Begins Test for Next Attempt," *New York Times* 29 January 1962: 45; and "Glenn Starts Tests for Orbital Flight," *New York Times* 12 February 1962: 19.

86. Finney, "Pilots Will Control," 1.

87. Witkin, "Training for Space," 3.

88. Manchester, "The Senseless Race," 67. For similar descriptions, see, for example, "Man in Space," 42; "Should Future Astronauts," 86; "Because We Want To," *Newsweek* 6 February 1961: 52; Finney, "What's Next," D12; Laurence, "Goals in Space," D9; and Plumb, "Robots Suggested," 29.

89. "Reaching," 55.

90. "Freedom's Flight," 56.

91. "Everything A-Okay," and "On Shepard's Trail," *Newsweek* 17 July 1961: 52. Also see, for example, "Flight Report," *Time* 16 June 1961: 72; "Men on the Moon," 61; Richard Witkin, "In Fine Condition," *New York Times* 6 May 1961: 1; and "Shepard's Ride," *New York Times* 7 May 1961: D1.

92. Finney, "Astronauts Can't," 25.

93. Every article on Glenn's flight refers to Glenn's active control. Many of the articles quote Glenn's or a NASA official's claim that the flight proved man is an active participant. John Finney, for example, wrote that Mercury officials are "now convinced that man should no longer be viewed as primarily a passenger in an automated vehicle, but rather should and can be used as an active pilot in controlling and navigating the spacecraft." "Glenn Feels Pilot Can Replace Much of Spaceship Automation," *New York Times* 23 February 1962: 1. Also see, for example, "The New Ocean," 22; "Race to the Moon," *Newsweek* 19 March 1962: 68; Finney, "Astronauts Give View," 15; and "Let Man Take Over," *New York Times* 25 February 1962: D10.

94. "On the Bazoo," 63.

95. John W. Finney, "Cooper Hailed in Capital," *New York Times* 22 May 1963: 20. For similar assessments, see, for example, Richard Witkin, "Dramatic Return," *New York Times* 17 May 1963: 18; "Great Gordo," 17, 19; "Water, Water," *Newsweek* 10 June 1963: 70; and Ben Kocivar, "The Collier Trophy Award," *Look* 22 October 1963: 142.

96. Robert A. Heinlein, "All Aboard the Gemini," *Popular Mechanics*

May 1963: 116. The media described all of the flights in this way. See, for example, "Saga of the Liberty Bell," 35; "A-Okay—Almost All the Way," *Newsweek* 31 July 1961: 20; "Rhymes with Hurrah," *Newsweek* 15 October 1962: 52; and "The Cruise of the Vostok," *Time* 21 April 1961: 50.

97. Gans, "The Messages," 43.

98. "Big Push to Outer Space," *Readers Digest* April 1959: 67. Also see "Man in Space," *Time* 5 September 1960: 43.

99. Bernard Lovell, "The Greatest Challenge to Man," *New York Times Magazine* 21 April 1963: 13.

100. "Space and Serendipity," *New York Times* 24 May 1961: 40. See also "The Cruise of the Vostok," 50.

101. Lovell, "The Greatest Challenge," 13. One can find other examples of the media focusing on a human's ability to respond to the unexpected. See, for example, "Lunar Soil Test Part of Mission," *New York Times* 31 July 1962: 12; Edwin Diamond, "That Moon Trip," 25; and Finney, "Astronauts Can't," 116.

102. Finney, "Astronauts Can't," 116.

103. "John Glenn: One Machine," 24.

104. "Great Gordo," 17. For a similar description, see page 18. Interestingly, *Newsweek* described Cooper's flight in terms identical to Glenn's. "On the nineteenth orbit, the machine faltered," the magazine stated, and Cooper "had to meet his test." "On the Bazoo," 61.

105. Moley, "What's Back of a Hero," 116.

106. Kocivar, "The Collier Trophy Award," 149.

107. Finney, "Astronauts Can't," 25.

108. Finney, "Astronauts Can't," 24.

109. Finney, "Astronauts Can't," 24.

110. "Let Man Take Over," D10.

111. "Let Man Take Over," 10.

112. "Let Man Take Over," 10.

113. "Let Man Take Over," 10. Also see Finney, "Race in Space," D8; "Space Goals," D1; and John W. Finney, "Space Debate Grows Sharper," *New York Times* 6 October 1963: D5.

114. "Nerveless?" 23.

115. "John Glenn: One Machine," 19–24.

116. "Future High Flier," *New York Times* 30 November 1961: 20.

117. Moley, "What's Back of a Hero," 116.

118. "The Flight of Friendship 7," *Newsweek* 29 January 1962: 72.

119. William Boot, "NASA and the Spellbound Press," 27.

4. *Life:* NASA's Mouthpiece in the Popular Media

1. David M. Berg, "Rhetoric, Reality, and Mass Media," *Quarterly Journal of Speech* 58 (1972): 255.

2. Researchers have found a tremendous similarity in the stories reported by the news media. Some observers propose that gatekeepers may

decide to include or exclude an item on the basis not only of its newsworthiness but also of the mix of news for that day. Gatekeepers' judgments may be influenced by whether "the 'minimum daily requirement' of that story topic has been met." Daniel Riffe, Brenda Ellis, Momo K. Rogers, Roger L. Van Ommeren, and Kieran A. Woodman, "Gatekeeping and the Network News Mix," *Journalism Quarterly* (Winter 1985): 321. See also Guido H. Stempel III, "Gatekeeping: The Mix of Topics and the Selection of Stories," *Journalism Quarterly* (Winter 1985): 791–815; and D. Charles Whitney and Lee B. Becker, "Keeping the Gates for Gatekeepers: The Effects of Wire News," *Journalism Quarterly* (Spring 1982): 60–65. For information on the agenda-setting function of the media, see Maxwell E. McCombs and Donald L. Shaw, "The Agenda-Setting Function of the Mass Media," *Public Opinion Quarterly* 36 (1972): 176–87; Donald L. Shaw and Maxwell E. McCombs, *The Emergence of American Political Issues: The Agenda-Setting Function of the Press* (St. Paul, Minn.: West Publishing, 1977); Robert D. McClure and Thomas E. Patterson, "Print vs. Network News," *Journal of Communication* 26 (Spring 1976): 23–28.

3. Lloyd F. Bitzer, "Rhetorical Public Communication," *Critical Studies in Mass Communication* 4 (1987): 426.

4. Bitzer, "Rhetorical Public Communication," 426.

5. John K. Jessup, "Henry R. Luce, 1898–1967: The Values That Shaped His Work," *Life* 10 March 1967: 30. Thomas Griffith proposed that in journalistic history Luce, "will be regarded as the great innovator of his time." *How True: A Skeptic's Guide to Believing the News* (Toronto: Little, Brown, 1974), 91.

6. Loudon Wainwright has written that "the highest echelons of Time, Inc., management regularly confused their journals with some nonexistent arm of government." *The Great American Magazine: An Inside History of "Life"* (New York: Knopf, 1986), 274.

7. David Halberstam, *The Powers That Be* (New York: Norton, 1977), 46. Luce himself stated that magazines could not be objective. See, for example, his comments in two speeches, "Causes, Causes!" and "Objectivity" in *The Ideas of Henry Luce*, ed. John K. Jessup (New York: Atheneum, 1969) (hereinafter cited as *The Ideas*), 56–57 and 71. W. A. Swanberg, *Luce and His Empire* (New York: Scribner's, 1972) is extremely critical of Luce's unacknowledged blending of fact and opinion, 141–42.

8. Swanberg, *Luce and His Empire*, 141. Halberstam concurs, labeling Luce a "true ideologue of the West," *The Powers*, 50.

9. "Journalism and Responsibility," 20 February 1953, in *The Ideas*, 80. In "National Purpose and Cold War," 28 June 1960, Luce asserted that the United States "must win the Cold War, and the sooner the better." He added, "Communism must be so stopped from spreading that men can confidently foresee its withering away." *The Ideas*, 132 and 133.

10. Curtis Prendergast and Geoffrey Colvin, *The World of Time, Inc.: The Intimate History of a Publishing Enterprise*, ed. Robert Lubar, vol. 3 (New York: Atheneum, 1986), 42. The magazine openly stated its commitment to the two aims in 1961. C. D. Jackson, *Life's* publisher, asked the rhetorical question, "Can a magazine presume to say that it will help win the Cold

War, help create a better America? It cannot presume otherwise." C. D. Jackson, "The Aim of LIFE," *Life* 2 June 1961: 1.

11. Loudon Wainwright, "He Hit That Keyhole in the Sky," *Life* 2 March 1962: 20–27; Lincoln Barnett, "The English Language," *Life* 2 March 1962: 74–83; James P. Wood commented that *Life* was "formulated to instruct as well as to attract, entertain, and amuse." He also points to the magazine's tendency to attract attention by publishing "sex pictures" and "gruesome pictures displaying slaughter, executions, and strewn corpses." *Magazines in the United States* (New York: Ronald Press, 1971), 215. Also see Luce's own comments about the provocative pictures that the magazine published. "Address to the Commissars," 30 April 1937 in *The Ideas*, 40.

12. Bayard Hooper, "The Champ Gate-Crasher," *Life* 2 March 1962: 10–12; "Who's the New Champ? She's Bing's Girl," *Life* 2 March 1962: 43–44; "A $70 Million Eccentric," *Life* 2 March 1962: 65–70; "Sam, the Man Behind the Colt," *Life* 2 March 1962: 57–60; "Double Debut in Paris Style," *Life* 2 March 1962; and "Waddling Whatzits," *Life* 2 March 1962: 94.

13. Dora Jane Hamblin, *That Was the Life* (New York: Norton, 1977), 18.

14. Robert T. Elson, *The World of Time, Inc.: The Intimate History of a Publishing Enterprise*, ed. Duncan Norton-Taylor, vol. 1 (New York: Atheneum, 1973), 423.

15. Hamblin, *That Was the Life*, 51.

16. Estelle Jussim, *Visual Communication and the Graphic Arts* (New York: Bowker, 1974), 8.

17. Maitland Edey, *Great Photographic Essays in "Life"* (New York: Little, Brown, 1978), 1.

18. Robert Rhode and Floyd McCall, *Press Photography: Reporting with a Camera* (New York: Macmillan, 1961), 194. Kenneth Burke makes a similar point when discussing identification through form and content. See *Counter-Statement* (Berkeley: University of California Press, 1968), 31, 124, 138.

19. Halberstam, *The Powers*, 60.

20. Halberstam, *The Powers*, 65.

21. "The Photograph and Good News," 2 September 1937, in *The Ideas*, 45.

22. Luce, "The Photograph," 44.

23. See, for example, "The Lonely Land That May Blow Up," *Life* 27 October 1961: 36–52; "Far Off War We Have Decided to Win," *Life* 16 March 1962: 37–39; Scot Leavitt, "Dilemma for U.S.: Clever Adversary, a Difficult Friend," *Life* 16 March 1962: 40–45. According to James Boylan, by 1955, American public relations operations had already implemented "a policy of uncritical support of Ngo Dinh Diem's client government." Boylan identifies Luce's magazines as a crucial participant in the effort. Whereas some publications began to adopt a more adversarial role in the 1960s, *Life* seems to have retained its partnership with the government. James Boylan, "Declaration of Independence," *Columbia Journalism Review* 25 (November/December 1986): 32.

24. Van Riper, *Glenn*, 147. See also Swenson, Grimwood, and Alexander, *This New Ocean*, 238; Sherrod, "The Selling of the Astronauts," 16–25.

25. Wainwright, *The Great American Magazine*, 261; the press, and NASA, said little about the wives' contract. "The Big Story," *Time* 24 August 1959, does reveal that the wives were "contract signatories," 38.

26. House, *1962 NASA Authorization*, pt. 1, 149.

27. *1962 NASA Authorization*, pt. 1, 149; Robert Sherrod offered the following comment about the memo allowing the astronauts to sell their personal stories: "With this memorandum, a Pandora's box was opened which spewed forth demons for the ensuing decade, involving not only NASA but also Congress, a large segment of the press and public, and the President of the United States." "The Selling of the Astronauts," 17.

28. Hamblin, *That Was the Life*, 18.

29. Wainwright, *The Great American Magazine*, 252.

30. Elson, *The World of Time*, vol. 1, 415. Prendergast and Colvin state that although *Life* had accounted for over half of corporate revenues throughout the 1950s, by 1959 its profits had "disappeared." In 1960, the magazine lost $1.1 million. Doubtless, Luce had patriotic motives for purchasing the astronauts' stories. It appears, however, that the magazine was economically motivated as well, hoping to increase slumping sales figures, *The World of Time*, vol. 3, 9.

31. Wainwright, *The Great American Magazine*, 252. Also see note 23.

32. Prendergast and Colvin, *The World of Time*, vol. 3, 9.

33. *Look* surpassed *Life* in sales in 1963 because *Life* abandoned the circulation race with *Look* in an attempt to upgrade the magazine and attract advertisers. See Prendergast and Colvin, *The World of Time*, vol. 3, 53. Figures in *The World Almanac and Book of Facts for 1962* (New York: New York World-Telegraph, 1963), 542, show *Look* with a slight sales lead. The figures, however, only account for sales until 30 September 1962. Those figures list the best-selling magazines as: (1) *Reader's Digest*, 11.3 million; (2) *TV Guide*, 7.4 million; (3) *McCall's*, 7.4 million; (4) *Ladies Home Journal*, 7 million; (5) *Look*, 6,854,000; (6) *Life*, 6,845,000. The following year, the figures available through 15 September 1963, show *Life* in fifth place, with 7,151,000 magazines sold, as compared with *Look*'s fourth place showing of 7,200,000 magazines sold. *The World Almanac and Book of Facts for 1963* (New York: New York World-Telegraph, 1964), 541.

34. Hamblin, *That Was the Life*, 34.

35. Halberstam, *The Powers*, 352–53.

36. Wainwright, *The Great American Magazine*, 263.

37. Diamond, *The Rise and Fall*, 88.

38. Finney, "U.S. Will Allow," 40.

39. Diamond, *The Rise and Fall*, 87.

40. Diamond, *The Rise and Fall*, 84.

41. Van Riper, *Glenn*, 147.

42. Wainwright, *The Great American Magazine*, 270.

43. Van Riper, *Glenn*, 146.

44. Wainwright, *The Great American Magazine*, 271.

45. Wainwright, *The Great American Magazine*, 263.

46. Kuo-jen Tsang, "News Photos in *Time* and *Newsweek*," *Journalism Quarterly* 61 (Autumn 1984): 578.

47. Wilson Hicks, *Words and Pictures* (New York: Arno, 1973), 5.

48. Susan Sontag, *On Photography* (New York: Farrar, Straus and Giroux, 1977), 17, 18.

49. Tsang, "News Photos," 578–84; C. E. Swanson, "What They Read in 130 Daily Newspapers," *Journalism Quarterly* 32 (1955): 411–21; and B. W. Woodburn, "Reader Interest in Newspaper Pictures," *Journalism Quarterly* 24 (September 1947): 197–201.

50. Sontag, *On Photography*, 5.

51. André Bazin, "The Ontology of the Photographic Image," in *Classic Essays on Photography*, ed. Alan Trachtenberg (New Haven, Conn.: Leete's Island Books, 1980), 241.

52. Paul Hightower, "A Study of the Messages in Depression-Era Photos," *Journalism Quarterly* 57 (Autumn 1980): 497.

53. Neil Postman, *Amusing Ourselves to Death* (New York: Penguin, 1985), 72–73.

54. Sontag, *On Photography*, 23.

55. Jean Berger and Jean Mohr, *Another Way of Telling* (New York: Pantheon, 1982), 92.

56. "World Will Be Ruled from Skies Above," *Life* 17 June 1963: 4.

57. John Dille, "Flight That Left Us Behind Tracked by Soviet Ships Off Our Coast," *Life* 14 August 1962: 25.

58. Dille, "Flight That Left Us," 25.

59. Paul Mandel, "Our Next Goal Man on the Moon," *Life* 27 April 1962: 85. See also "World Will Be Ruled," 4, and "Space: An American Necessity," *Life* 30 November 1959: 36.

60. "Space: An American Necessity," 36. Paul Mandel cited the "technological benefits," the advance of "our technology" in fields "far removed from space travel," and our improved "knowledge of basic science" as reasons for going to the moon. "Our Next Goal," 85.

61. "Hitching on to the Infinite," *Life* 15 June 1962: 4.

62. Dille, "Flight That Left Us," 25.

63. "Space: An American Necessity," 36.

64. Mandel, "Our Next Goal," 85.

65. "Stake in Space: Our Survival," *Life* 24 August 1962: 4. In the same editorial, the author proclaimed that the Soviets "always related the means of power to political ends." Americans could be certain that the Soviets would not "shrink from using their space power to dominate terrestrial affairs."

66. "Space: An American Necessity," 36. Like the administration, *Life* also proposed that America's space objectives were peaceful. "American ideas include the maintenance of space as a peaceful and ordered domain of mankind." "Liftoff and Uplift for the U.S.; a World's Hope," *Life* 2 March 1962: 4.

67. "The Astronauts—Ready to Make History," *Life* 14 September 1959: 26. For a similar description, see " 'Life' Is With It in a Far-Out Era," *Life* 14 September 1959: 2, and Loudon S. Wainwright, "The Three Chosen for the First Space Ride," *Life* 3 March 1961: 24.

68. Loudon Wainwright, *The Astronauts: Pioneers in Space* (New York: Golden Press, 1961), 9 (hereinafter cited as *The Astronauts*). Wainwright

attempted to compare the astronauts with the pioneers whenever possible. He entitled his chapters, "Seven Pioneers" and "The Space Frontier."

69. John Dille, in *We Seven* (New York: Simon and Schuster, 1962), 3.

70. Dille, in *We Seven*, 18. Paul Mandel also described the moon shot as the "greatest of adventures." "Our Next Goal," 62.

71. "A Costly Trip to the Moon," *Life* 16 June 1961: 54.

72. Mandel, "Our Next Goal," 85.

73. "World Will Be Ruled," 4.

74. "Space: An American Necessity," 36. Thompson returns to this theme often. He justified the program, for example, by pointing to "man's eternal thirst for knowledge, the call of human adventure." At times, he tried to make the moon shot the ultimate test of America's greatness: "Now the U.S. can be foremost in another and greater adventure—or abdicate its own greatness by not doing enough." "World Will Be Ruled," 4.

75. "The Astronauts—Ready to Make History," 26. Dille, in *We Seven*, 21.

76. Wainwright, *The Astronauts*, 9.

77. "World Will Be Ruled," 4.

78. Dille, in *We Seven*, 4. See also Mandel, "Our Next Goal," 70.

79. Mandel, "Our Next Goal," 85. The magazine highlighted the danger of the missions in its first article. "If he [the astronaut] survives, he will become the heroic symbol of a historic triumph." "The Astronauts—Ready," 26.

80. Mandel, "Our Next Goal," 81.

81. "Stake in Space," 4.

82. "Soviet Trader Returns from Out of This World," *Life* 21 April 1961: 24.

83. The magazine provided perhaps its only humanizing picture of the astronauts early in the program. Despite the astronauts' "extraordinary qualifications," *Life* stated, they have "many of the preoccupations, and even small weaknesses, of more ordinary men. Two of the four cigarette smokers in the group are trying—so far unsuccessfully—to stop. Two others are worried about their weight. They are concerned about the condition of the grass in their yards and proper schooling for their children." "The Astronauts—Ready to Make History," 27. These "weaknesses," however, are never mentioned again.

84. Van Riper, *Glenn*, 147.

85. Van Riper, *Glenn*, 146.

86. "The Astronauts—Ready," 27.

87. Wainwright, *The Astronauts*, 92.

88. Dille, in *We Seven*, 4.

89. Dille, in *We Seven*, 4.

90. Dille, in *We Seven*, 6 and 4.

91. Wainwright, *The Great American Magazine*, 270.

92. Wainwright, *The Great American Magazine*, 270.

93. Wainwright, "The Three Chosen," 30. Some articles made references to the flight that must have left questions about how much control Shepard exerted. One article referred to Shepard and "the capsule he rode in." "Emotions of the Nation Ride in Astronaut's Capsule," *Life* 12 May 1961:

19. Another article referred to the "Redstone rides." "Hero's Welcome and Week Off," *Life* 19 May 1961: 32.

94. "Details of a Scientific and a Flawless Flight," *Life* 12 May 1961: 22.

95. "Details of a Scientific," 22 and 23. For a similar description, see "Emotions of the Nation," 19.

96. Wainwright, "The Three Chosen," 28.

97. Dille, in *We Seven*, 10.

98. Dille, in *We Seven*, 9.

99. Wainwright, "The Three Chosen," 27.

100. "A Man Marked to Do Great Things," *Life* 2 February 1962: 28.

101. "A Man Marked," 24.

102. "Transcript of Glenn's Talk," *New York Times* 4 March 1962: 4.

103. Wainwright, "The Three Chosen," 27.

104. Wainwright, "The Three Chosen," 27.

105. Wainwright, "The Three Chosen," 27.

106. Dora Jane Hamblin, "Applause, Tears and Laughter, and the Emotions of a Long-Ago Fourth of July," *Life* 9 March 1962: 35.

107. "Liftoff and Uplift," 4.

108. Hamblin, "Applause," 34. She describes Glenn and his speech in patriotic terms. She said, for example, that Glenn held Congress as "spellbound as Boy Scouts hearing how the Scoutmaster killed a nine-foot snake." She also said the speech "touched an emotion as fundamental as the memory of long-ago Fourths of July and the uncomplicated passion of a schoolroom Pledge of Allegiance." Ibid., 34.

109. Hamblin, "Applause," 35. See also "Hero's Words to Cherish," *Life* 9 March 1962: 4.

110. "He Hit That Keyhole," 20.

111. Paul Mandel, "The Ominous Failures That Haunted Friendship's Flight," *Life* 2 March 1962: 39. Mandel also stated that after the failure, Glenn spent "90% of his time flying the capsule."

112. Mandel, "The Ominous," 39.

113. Loudon Wainwright, "Comes a Quiet Man to Ride Aurora 7," *Life* 18 May 1962: 33.

114. Wainwright, "Comes a Quiet Man," 33.

115. Wainwright, "Comes a Quiet Man," 34.

116. Wainwright, *The Great American Magazine*, 276.

117. Wainwright, *The Great American Magazine*, 276–77.

118. Wainwright, *The Great American Magazine*, 276.

119. John Dille, "At the End of a Great Flight, Big Bull's-Eye," *Life* 12 October 1962: 50.

120. "This Is a Victory for the Human Spirit," *Life* 24 May 1963: 31.

121. "This Is a Victory," 31.

122. *Book Review Digest*, ed. Dorothy P. Davidson (New York: H. W. Wilson, 1963), describes the book as one for readers "ten and up," 464.

123. Richard Witkin, "Pioneers in Space," *New York Times Book Reviews* 18 November 1962: 16.

124. Norbert Bernstein, "Book Reviews," *Library Journal* 1 November 1962: 4034.

125. Witkin, "Pioneers in Space," 16.

126. R. C. Cowen, "Ordinary Supermen," *Christian Science Monitor* 15 November 1962: 14b.

127. Bernstein, "Book Reviews," 4034.

128. See, for example, Virgil Grissom, "Countdown," in *The Astronauts* 48, 49, 52; Gordon Cooper, "First Rocket We Will Ride," *Life* 3 October 1960: 78, and "Our First Step," in *The Astronauts*, 18, 19; John Glenn, "Readying Mind and Body," in *The Astronauts*, 27, 29, 33; Walter Schirra, "Our First Step," in *The Astronauts*, 56, 59; Scott Carpenter, "In Orbit," in *The Astronauts*, 64.

129. See the comments of Virgil Grissom, Alan Shepard, John Glenn, Walter Schirra, Gordon Cooper, and Scott Carpenter in *We Seven*, 159, 162, 223–26, 291–97, 25, 31, 177, 262, 326, 351, 61, 92, 94, 54, 265, 353. Slayton and Grissom state that when they entered the program, they thought that the whole program was a stunt but that after they entered the program, they realized that only real test pilots could handle it (71 and 55).

130. With few exceptions, the astronauts referred to their vessel as a "capsule" in *The Astronauts*. For descriptions of their vessels as "rockets," see notes 128–29.

131. John Glenn, "We're Going Places No One Has Ever Travelled in a Craft No One's Flown," *Life* 27 January 1961: 38.

132. See notes 128–29.

133. Grissom, "Countdown," 53, and Gordon Cooper, "Our First Step," in *The Astronauts*, 19.

134. See, for example, Glenn and Grissom, in *We Seven*, 351 and 55.

135. See Glenn and Shepard, in *We Seven*, 24 and 164.

136. For comments on the unknown environment, see *We Seven*, 45, 153, 162, 179; for references to the pioneers and the Old West, see *We Seven*, 37, 59, 61, 67, 351, 354.

137. The wives were "contract signatories" with their husbands in the original *Life* contract. "The Big Story," *Time* 24 August 1959: 38.

138. Diamond, *The Rise and Fall*, 87.

139. Anna Glenn, "Seven Brave Women Behind the Astronauts," *Life* 21 September 1959: 142.

140. Rene Carpenter, "There Are No Dark Feelings," *Life* 21 September 1959: 146; see also Trudy Cooper, "I Want to Watch It Go," *Life* 21 September 1959: 157.

141. Anna Glenn, "Seven Brave Women," 142.

142. Trudy Cooper, "I Want to Watch," 157.

143. Jo Schirra, "Maybe I've Been Lucky," *Life* 21 September 1959: 158.

144. Marjorie Slayton, "I Have Never Been Nervous," *Life* 21 September 1959: 160.

145. Louise Shepard, "Just Go Right Ahead," *Life* 21 September 1959: 150.

146. Louise Shepard, "Just Go Right," 150.

147. Trudy Cooper, "I Want to Watch," 157.

148. Betty Grissom, "Nothing So Important As I Love You," *Life* 28 July 1961: 29.

149. Trudy Cooper, "I Felt Calm," *Life* 31 May 1963: 36.

150. Trudy Cooper, "I Felt Calm," 36.

151. Prendergast and Colvin, *The World of Time*, vol. 3, 39.
152. "The Astronaut's Story Will Appear Only in LIFE," *Life* 24 August 1959: 98.
153. Hamblin, *That Was the Life*, 51.
154. "Space Voyagers Rarin' to Orbit," *Life* 20 April 1959: 24–25.
155. See "The Astronauts—Ready," 26–27 and the accompanying articles by the astronauts.
156. "He Hit That Keyhole," 22–23.
157. "He Hit That Keyhole," 24–26. See also Loudon Wainwright, "For Those Who Cared Most, the Long Watch at Home," *Life* 27 September 1963: 70–90.
158. "Head Over Heels," 70–90.
159. Donald Slayton, "We Believe They Should Leave the Flying to Us," *Life* 27 September 1963: 90.
160. William F. Lewis, "Telling America's Story: Narrative Form and the Reagan Presidency," *Quarterly Journal of Speech* 73 (1987): 289.
161. Wainwright, *The Great American Magazine*, 273–74.
162. Letter from Edward Thompson to James Webb, 30 March 1962, NASA HO.
163. Letter from James Webb to Edward Thompson, 23 April 1962, NASA HO.
164. The seven original Mercury astronauts and the nine new Apollo astronauts signed a four-year, $1,040,000 contract with Field Enterprises Educational Corporation (publisher of *World Book Encyclopedia*) and *Life* magazine. The contract also created an option for a four-year renewal of the contract. Field Enterprises paid each astronaut $10,000 per year for newspaper syndication and book rights. *Life* paid each astronaut $6,500 per year for magazine rights. "Astronauts Sell Personal Stories," *New York Times* 18 September 1963: 15. See also Sherrod, "The Selling of the Astronauts," 21. Prendergast and Colvin wrote that Vice President Lyndon B. Johnson, assigned to oversee the astronauts' new contract in 1962, "privately sought out *Life's* guidance." Edward Thompson supplied Johnson with "a talking brief" to use in justifying a new contract with *Life*, *The World of Time*, vol. 3, 60.
165. Van Riper, *Glenn*, 148.
166. Van Riper, *Glenn*, 148.
167. Van Riper, *Glenn*, 148.
168. Memorandum for Deputy Administrator Hugh Dryden for Walter L. Lingle, Jr., Acting Assistant Administrator for Public Affairs, "Policy Concerning the Astronauts," 1 August 1962, 2, NASA HO.
169. "Policy Concerning the Astronauts," 1.
170. Wainwright, *The Great American Magazine*, 263.

5. Congressional Space Committees: Overseers or Advocates?

1. Woodrow Wilson, *Congressional Government* (Baltimore: Johns Hopkins University Press, 1981), 24. More current sources demonstrate that

little has changed. See William J. Keefe and Morris S. Ogul, *The American Legislative Process: Congress and the States* (Englewood Cliffs, N.J.: Prentice-Hall, 1993), 169–70; and Steven S. Smith and Christopher J. Deering, *Committees in Congress* (Washington, D.C.: Congressional Quarterly Press, 1984), 271–72.

2. John R. Fitzpatrick, "Congressional Debating," *Quarterly Journal of Speech* 27 (1941): 251–52.

3. House, *Discussion of Soviet Man-in-Space Shot*, 7, 8. For similar testimony, see House, *1962 NASA Authorization*, pt. 1, 145–46, 364–65; pt. 2, 631, 634–35, 644, 799–800; pt. 3, 1047.

4. House, *1962 NASA Authorization*, pt. 2, 826–27, 619, 800, 828.

5. House, *Discussion of Man-in-Space Shot*, 13–14.

6. House, *1963 NASA Authorization*, pt. 2, 418, 419, 991. Also see the testimony of Emilio Q. Daddario, in House, *1964 NASA Authorization*, pt. 2 (a), 930.

7. House, *1963 NASA Authorization*, pt. 2, 441, 451, 418.

8. House, *1962 NASA Authorization*, 18.

9. House, *1962 NASA Authorization*, 82.

10. Senate, *Independent Offices Appropriations, 1964*, 1520, 1527, 1528, and 1619.

11. House, *1962 NASA Authorization*, 376.

12. House, *Discussion of Man-in-Space Shot*, 23.

13. House, *Discussion of Man-in-Space Shot*, 19, and House, *1962 NASA Authorization*, pt. 1, 371; pt. 2, 827.

14. House, *1962 NASA Authorization*, pt. 2, 718–19 and 653.

15. For additional comments supporting an accelerated space program, see the testimony of Emilio Q. Daddario, in House, *Discussion of Man-in-Space Shot*, 22; William F. Ryan, in House, *1962 NASA Authorization*, pt. 2, 827; and Ken Hechler, pt. 1, 369, 170; pt. 2, 613.

16. Senate, *Independent Offices Appropriations, 1962*, 653 and 651.

17. Senate, *Independent Offices Appropriations, 1962*, 655 and 654. Representatives Fulton and Chenoweth addressed the military issue in 1961 as well. See House, *1962 NASA Authorization*, pt. 2, 634; pt. 3, 1052.

18. Ken Hechler, *The Endless Space Frontier: A History of the House Committee on Science and Astronautics, 1959–1978* (San Diego: American Astronautical Society, 1982), 167. He offers a similar comment during the 1964 authorization hearings. See House, *1964 NASA Authorization*, pt. 2 (a), 620. Also see pt. 2, 435.

19. House, *1964 NASA Authorization*, pt. 2 (a), 620. For additional testimony, see 254 and 408.

20. House, *1963 NASA Authorization*, pt. 2, 617, 619.

21. House, *1964 NASA Authorization*, pt. 1, 31–32.

22. House, *1964 NASA Authorization*, pt. 2 (a), 632. For additional testimony, see 567.

23. House, *1964 NASA Authorization*, pt. 2 (a), 624 and 555.

24. U.S. Congress, Senate, Committee on Aeronautical and Space Sciences, *Scientists' Testimony on Space Goals*, 88th Cong., 1st sess., 1963, 104 and 133 (hereinafter cited as Senate, *Scientists' Testimony*).

25. Senate, *Scientists' Testimony*, 173. For other comments, see Senator Margaret Chase Smith, 133, and Senator Edmondson, 20.

26. House, *1964 NASA Authorization*, pt. 2 (a), 621.

27. Senate, *Scientists' Testimony*, 103.

28. House, *1964 NASA Authorization*, pt. 1, 98.

29. House, *1962 NASA Authorization*, pt. 2, 672; pt. 1, 375; and House, *Discussion of Man-in-Space Shot*, 27.

30. House, *1962 NASA Authorization*, pt. 2, 829. Also see House, *1963 NASA Authorization*, pt. 1, 11.

31. House, *1962 NASA Authorization*, pt. 1, 380; pt. 2, 1088; and House, *Discussion of Man-in-Space Shot*, 3, 4.

32. Senate, *Independent Offices Appropriations, 1964*, 1569.

33. House, *Discussion of Man-in-Space Shot*, 12.

34. House, *1962 NASA Authorization*, pt. 3, 1060. For additional remarks on the prestige value of the moon shot, see the testimony of Albert King, House, *Discussion of Man-in-Space Shot*, 26; Robert Kerr, Stuart Symington, Stephen M. Young, in Senate, *Orbital Flight*, 1, 15, and 19.

35. Senate, *Orbital Flight*, 10. Also see the questioning of Senator Cannon, 23.

36. House, *1963 NASA Authorization*, pt. 1, 146. Also see the questioning of Chairman Overton Brooks, 77 and 79.

37. Senate, *Orbital Flight*, 25.

38. House, *Independent Offices Appropriations for 1962*, 653. Also see House, *1962 NASA Authorization*, pt. 1, 378; pt. 3, 1052.

39. Senate, *Orbital Flight*, 16. Also see the comments of Chairman Kerr and Senator Howard W. Cannon, 1 and 21–22. Overton Brooks made similar comments, House, 1963 *NASA Authorization*, 1.

40. Senate, *Scientists' Testimony*, 81. Also see Chairman Albert Thomas, House, *Independent Offices Appropriations for 1963*, pt. 3, 81.

41. House, *1963 NASA Authorization*, pt. 2, 718, 884.

42. U.S. Congress, House, Committee on Appropriations, *Second Supplemental Appropriation Bill, 1962*, 87th Cong., 2d sess., 1962, 398.

43. U.S. Congress, House, Committee on Appropriations, *Independent Offices Appropriations for 1964*, 88th Cong., 1st sess., 1963, pt. 3, 98. Also see, for example, the testimony of Harold C. Ostertag, in House, *Independent Offices Appropriations, 1963*, pt. 3, 887; Edgar Chenoweth, in House, *1962 NASA Authorization*, 377; and James C. Corman, pt. 2, 829.

44. Senate, *Scientists' Testimony*, 22.

45. Senate, *Scientists' Testimony*, 143.

46. Senate, *Scientists' Testimony*, 109, 106.

47. Senate, *Scientists' Testimony*, 78. For additional support of the scientific aspects of the program, see, for example, Dr. H. H. Hess, 181, and Dr. Martin Schwarzchild, 163.

48. Senate, *Scientists' Testimony*, 51.

49. Senate, *Scientists' Testimony*, 63, 66.

50. Senate, *Scientists' Testimony*, 5.

51. Senate, *Scientists' Testimony*, 3–4.

52. Senate, *Scientists' Testimony*, 6.

53. For a discussion of technical argument, see G. Thomas Goodnight, "The Personal, Technical, and Public Spheres," 214–27; Farrell, "Knowledge, Consensus," 1–14; and Thomas B. Farrell and Thomas Goodnight, "Accidental Rhetoric," 271–300.

54. Senate, *Scientists' Testimony*, 106.

55. Senate, *Scientists' Testimony*, 28, 31, 32.

56. Senate, *Scientists' Testimony*, 32. Although Ramo did not believe manned space flight was justified before America had exhausted the capabilities of unmanned exploration, he acknowledged the prestige value of manned flights. When he was asked why men were put in the space capsules, he responded as follows: "There was a good deal more glamour built up around the question than cold logic would dictate" (13).

57. Senate, *Scientists' Testimony*, 139.

58. Senate, *Scientists' Testimony*, 55, 51. For similar testimony, see, for example, Dr. Frederick Seitz, 88; Dr. Schwarzchild, 163, and Dr. Berkner, 108.

59. Senate, *Scientists' Testimony*, 98. For a similar opinion, See Dr. Berkner, 107, 128.

60. Senate, *Scientists' Testimony*, 58. For similar testimony, see Dr. Kush, 72, and Dr. Pittendrigh, 86.

61. Senate, *Scientists' Testimony*, 15, 21.

62. See Dr. Ramo, and Dr. H. H. Hess, in Senate, *Scientists' Testimony*, 26–27, 32, 34, and 184.

63. See Fisher, "Narration as Human Communication," 1–22.

64. Senate, *Scientists' Testimony*, 89.

65. Senate, *Scientists' Testimony*, 76.

66. Senate, *Scientists' Testimony*, 67. For an additional comment, see 64.

67. Senate, *Scientists' Testimony*, 84. For an additional comment, see 75.

68. Senate, *Scientists' Testimony*, 164, 174, and 162. For similar comments, see also Dr. Seitz, 93; Dr. H. H. Hess, 179–80; Dr. Urey, 51, 53, 57; Dr. Berkner, 108; and Dr. DuBridge, 138–39. Abelson attacked the comparison between space exploration and the voyage of Columbus, arguing that Columbus sought economic returns. With our large microscopes, Abelson added, we can tell that the moon holds "no objects of economic value" (5). Yet even Abelson cites exploration as a fundamental human urge: "Man will explore the unknown, that is his nature" (23).

69. John W. Finney, "Biologist Backs Space Plan Foes," *New York Times* 9 June 1963: 21.

70. Senate, *Scientists' Testimony*, 73–74.

71. See McDougall, *The Heavens*, 315–16.

72. Amitai Etzioni, "When Scientists Testify," *Bulletin of the Atomic Scientists* 20 October 1964: 25.

73. Etzioni, "When Scientists Testify," 24.

74. James R. Kerr, "Congressmen as Overseers: Surveillance of the Space Program" (Ph.D. diss., Stanford University, 1963), 106.

75. Hechler, *The Endless Space*, 161.

76. Hechler, *The Endless Space*, 162.

77. Kerr, "Congressmen as Overseers," 106.

78. For a discussion of the relationships between space committee leaders and the administration, see McDougall, *The Heavens*, 361–88.

79. U.S. Congress, House, Committee of the Whole House on the State of the Union, *Authorizing Appropriations to the National Aeronautics and Space Administration: Report to Accompany H.R. 7500*, 88th Cong., 1st sess., 1963, H. Rept. 591, 192 (hereinafter House, *Report Accompanying H.R. 7500*).

80. U.S. Congress, House, Committee of the Whole House on the State of the Union, *Authorizing Appropriations to the National Aeronautics and Space Administration: Report to Accompany H.R. 6874* 87th Cong., 1st sess., 1961, H. Rept. 391, 33–34 (hereinafter *Report Accompanying H.R. 6874*).

81. House, *1962 NASA Authorization*, pt. 1, 156–57, and pt. 2, 790–91. Also see House, *Independent Office Appropriations, 1962*, pt. 2, 1193, 1228, and 1232–33.

82. House, *Report Accompanying H.R. 6874*, 89.

83. House, *Report Accompanying H.R. 7500*, 201.

84. House, *Report Accompanying H.R. 7500*, 206.

85. U.S. Congress, Senate, Committee on Aeronautical and Space Sciences, *Authorizing Appropriations to the National Aeronautics and Space Administration: Report to Accompany H.R. 7500*, 88th Cong., 1st sess., 1963, H. Rept. 385, 348.

6. Justificatory Rhetoric: Floor Debates Concerning Project Apollo

1. Fitzpatrick, "Congressional Debating," 252. More recent sources propose that little has changed in the past fifty years. See Keefe and Ogul, *The American Legislative Process*, 169–70; and Steven S. Smith and Christopher J. Deering, *Committees in Congress* (Washington, D.C.: Congressional Quarterly Press, 1984), 271–72.

2. Miller, *Cong. Rec.* 23 May 1962: 9049.

3. Miller, *Cong. Rec.* 10 October 1963: 19265.

4. Kerr, "Congressmen as Overseers," 421.

5. Miller, *Cong. Rec.* 24 May 1961: 8839 and 8835.

6. Gross, *Cong. Rec.* 23 May 1962: 9095.

7. *Cong. Rec.* 24 May 1963: 8203.

8. Fulbright, *Cong. Rec.* 17 October 1963: 19763.

9. Ernest Gruening, *Cong. Rec.* 29 October 1963: 22441.

10. Clark, *Cong. Rec.* 8 August 1963: 14565.

11. Wyman, *Cong. Rec.* 10 October 1963: 19254. Critic Thomas Pelly offered a similar comment, proposing that he favored a "stretchout of the Apollo manned lunar landing program," which would save a great deal of money without harming the overall program. *Cong. Rec.* 1 August 1963: 13881.

12. Fulbright, *Cong. Rec.* 19 November 1962: 22363. Later in the same

debate, he proclaimed the following: "All I am trying to do is to slow down the pace of the program." "The pace of the program is not essential." *Cong. Rec.* 19 November 1963: 22370. For an identical comment, see *Cong. Rec.* 20 November 1963: 22451.

13. "Expanded Space Effort Viewed as Cold War Need, Poll Shows," *Aviation Week and Space Technology* 5 February 1962: 34.

14. Albert Brooks, *Cong. Rec.* 24 May 1961: 8828.

15. J. Edward Roush, *Cong. Rec.* 1 August 1963: 13882.

16. H. R. Gross, *Cong. Rec.* 1 August 1963: 13887. For similar comments on checking communism, see Edward J. Gurney, 10 October 1963: 19242, and Richard L. Roudebush, 24 May 1961: 8837.

17. Clinton Anderson, *Cong. Rec.* 13 April 1961: 5875, and *Cong. Rec.* 27 May 1963: 9500.

18. Olin Teague, *Cong. Rec.* 1 August 1963: 13855. Also see Charles Q. Wilson, 13 June 1963: 10856; J. Edgar Chenoweth, 1 August 1963: 13872; Emilio Daddario, 1 August 1963: 13878; Stephen Young, 22 March 1961: 4475; Stuart Symington, 19 November 1963: 22361; Joseph E. Karth, 24 May 1961: 8836; and Thomas J. Dodd, 29 August 1962: 17510–11.

19. J. William Fulbright, *Cong. Rec.* 19 November 1963: 22368. For other criticism of the prestige value, see Gordon Allott, 28 June 1961: 11627, and Richard Roudebush, 23 May 1962: 9087.

20. Joseph Clark, *Cong. Rec.* 9 August 1963: 14702. For a similar depiction, see Joe Skubitz, 10 October 1963: 19256.

21. Thomas M. Pelly, *Cong. Rec.* 7 June 1961: 9675.

22. James D. Weaver, *Cong. Rec.* 28 October 1963: 20385.

23. Teague, *Cong. Rec.* 1 August 1963: 13854.

24. Teague, *Cong. Rec.* 10 October 1963: 19241.

25. See, for example, Emilio Daddario, 23 May 1962: 9079; James Fulton, 24 May 1961: 8831; Albert Brooks, 24 May 1961: 8835; Albert Thomas, 10 October 1963: 19241; George Miller, 23 May 1962: 9051; Clinton Anderson, 19 November 1963: 22369; and Stuart Symington, 19 November 1963: 22361.

26. William Proxmire, *Cong. Rec.* 20 August 1962: 17106.

27. Proxmire, *Cong. Rec.* 10 August 1962: 13092.

28. Symington, *Cong. Rec.* 19 November 1963: 22361–62.

29. George Miller, *Cong. Rec.* 6 September 1962: 18671.

30. Representatives Richard Roudebush (Ind.), Thomas Pelly (Wash.), Donald Rumsfeld (Ill.), James D. Weaver (Pa.), Edward J. Gurney (Fla.), and John W. Wydler (N.Y.) called for greater emphasis on military exploitation of inner space. See House, *Report Accompanying H.R. 7500*, 201–05.

31. Republican congressmen Joe Skubitz (Kans.), J. Edgar Chenoweth (Colo.), H. R. Gross (Iowa), and Louis Wyman (N.H.) joined the above-mentioned group in pushing inner space during the floor debates. See, for example, Edward Gurney, 1 August 1963: 13889; Louis Wyman, 10 October 1963: 19252; Joe Skubitz, 10 October 1963: 19256; Thomas Pelly, 7 June 1961: 9675; Gordon Allott, 28 June 1961: 11629; Edgar Chenoweth, 1 August 1963: 13874; and Edward Roush, 1 August 1963: 13883 and 13884.

32. Weaver, *Cong. Rec.* 10 October 1963: 19240.

33. Weaver, *Cong. Rec.* 29 November 1963: 20385.
34. Donald Rumsfeld, *Cong. Rec.* 10 October 1963: 19256.
35. Daddario, *Cong. Rec.* 23 May 1962: 9079.
36. Margaret Chase Smith, *Cong. Rec.* 8 May 1963: 7959.
37. Fulton, *Cong. Rec.* 1 August 1963: 13859.
38. Miller, *Cong. Rec.* 23 May 1962: 9049. Also see, for example, Warren Magnuson, 19 November 1963: 22367; Olin Teague, 23 May 1962: 9064; Joseph E. Karth, 23 May 1962: 9075.
39. Thomas Pelly, *Cong. Rec.* 1 August 1963: 13880–81.
40. Fulbright, *Cong. Rec.* 19 November 1963: 22368.
41. Ken Hechler, *Cong. Rec.* 10 October 1963: 19258, and Edgar Chenoweth, *Cong. Rec.* 1 August 1963: 13873.
42. Anderson, *Cong. Rec.* 27 May 1963: 9500.
43. Teague, *Cong. Rec.* 23 May 1962: 9064. Also see, for instance, James Fulton, 17 October 1963: 19764, and Victor Anfuso, 23 May 1962: 9073.
44. Brooks, *Cong. Rec.* 24 May 1961: 8828.
45. Teague, *Cong. Rec.* 1 August 1963: 13855. Also see, for example, Carl Albert, 23 May 1962: 9092; James Fulton, 1 August 1963: 13859; George Miller, 23 May 1962: 9050; Albert King, 23 May 1962: 9085; Clinton Anderson, 27 May 1963: 9500; and Katharine St. George, 1 August 1963: 13853.
46. Fulbright, *Cong. Rec.* 17 October 1963: 19764, and 19 November 1963: 22371.
47. Louis C. Wyman, *Cong. Rec.* 10 October 1963: 19238.
48. Proxmire, *Cong. Rec.* 10 July 1962: 13090–97. For other criticism of the economic aspects, see Joseph Clark, 19 November 1963: 22367; John J. Rhodes, 30 July 1962: 14979; Thomas Pelly, 1 August 1963: 13880; Frank Lausche, 9 August 1963: 14693; Howard Cannon, 10 October 1963: 19226; and Carl T. Curtis, 9 August 1963: 14699.
49. James Fulton, *Cong. Rec.* 1 August 1963: 13859.
50. Teague, *Cong. Rec.* 23 May 1962: 9064. Also see, for example, Emilio Daddario, 1 August 1963: 13878; Carl Albert, 23 May 1962: 9092; and George Miller, 1 August 1963: 13862.
51. Fulbright, *Cong. Rec.* 17 October 1963: 19764, and Proxmire, 11 August 1962: 13252.
52. Teague, *Cong. Rec.* 1 August 1963: 13855.
53. Anderson, *Cong. Rec.* 27 May 1963: 9501.
54. Walter Riehlman, *Cong. Rec.* 1 August 1963: 13870.
55. John Sparkman, *Cong. Rec.* 31 October 1963: 20792.
56. Teague, *Cong. Rec.* 23 May 1962: 9077.
57. Chester E. Merrow, *Cong. Rec.* 8 May 1961: 7554.
58. Thomas J. Lane, *Cong. Rec.* 11 May 1961: 8831.
59. Karl Mundt, *Cong. Rec.* 8 May 1961: 7468.
60. Teague, *Cong. Rec.* 24 May 1961: 8844.
61. Alan Bible, *Cong. Rec.* 8 March 1962: 3608.
62. Harold D. Donohue, *Cong. Rec.* 20 February 1962: 2611.
63. Carl Albert, *Cong. Rec.* 20 February 1962: 2611.

64. Fred Schwengel, *Cong. Rec.* 26 February 1962: 2956.
65. Ralph Yarborough, *Cong. Rec.* 21 February 1962: 2741, and Mike Mansfield, *Cong. Rec.* 26 February 1962: 2891.
66. John M. Butler, *Cong. Rec.* 22 March 1962: 4865.
67. Robert Kerr, *Cong. Rec.* 28 June 1961: 11627. Also see James Fulton, 28 June 1961: 11625, and Joseph Karth, 1 August 1963: 13871.
68. Richard Fulton, *Cong. Rec.* 1 August 1963: 13906.
69. Lane, *Cong. Rec.* 11 May 1961: 7889.
70. Roush, *Cong. Rec.* 21 February 1962: 2682.
71. Garner E. Shriver, *Cong. Rec.* 21 May 1963: 9169.
72. Proxmire, *Cong. Rec.* 10 July 1962: 13089.
73. Gordon Allott, *Cong. Rec.* 17 May 1962: 8722.
74. Ben F. Jensen, *Cong. Rec.* 23 May 1962: 3045.
75. John Lindsay, *Cong. Rec.* 23 May 1962: 9061.
76. Maurine B. Neuberger, *Cong. Rec.* 20 November 1963: 22441.
77. Fulbright, *Cong. Rec.* 17 October 1963: 19764.
78. Lewis, "Telling America's Story," 289.

Conclusion

1. Arthur C. Clarke, "Space Flight and the Spirit of Man," *Reader's Digest* February 1962: 78.
2. Seth Payne and William D. Marbach, "Science Is Still Waiting for Lift-off," *Business Week* 17 October 1988: 50–51.
3. Boot, "NASA and the Spellbound Press," 27.
4. "Remarks on the 20th Anniversary of the Apollo Moon Landing," 20 July 1989, *Weekly Compilation of Presidential Documents* (Washington, D.C.: GPO, 1989), 1129.
5. William J. Cook, "The New Frontiers," *U.S. News and World Report* 26 September 1988: 50. Also see, for example, Leon Jaroff, "Onward to Mars," *Time* 18 July 1988: 46–53; Stanley N. Wellborn, "Man's Inevitable Trip to Mars," *U.S. News and World Report* 3 August 1987: 46–48; and "The Case for Space," *New Republic* 27 June 1988: 7–9.
6. Cook, "The New Frontiers," 50, 52.
7. See, for example, Carl Sagan, "Why Send Humans to Mars?" *Issues in Science and Technology* 7 (Spring 1991): 80–85; Carl Sagan, "The Case for Mars," *Discover* September 1984: 26; Carl Sagan, "It's Time to Go to Mars," *New York Times* 23 January 1987: A27; and Carl Sagan, "To Mars," *Aviation Week and Space Technology* 8 December 1986: 11.
8. McDougall, *The Heavens*, 305.
9. McDougall, *The Heavens*, 362.
10. Lyndon Baines Johnson, *The Vantage Point: Perspectives of the Presidency, 1963–1969* (New York: Holt, 1971), 285.
11. Johnson, *The Vantage Point*, 286.
12. Johnson, *The Vantage Point*, 285.

Bibliography

Books

Alexander, Tom. *Project Apollo: Man to the Moon.* New York: Harper, 1964.

Baugham, James L. *Henry R. Luce and the Rise of the American News Media.* Boston: Twayne, 1987.

Bazin, André. "The Ontology of the Photographic Image." In *Classic Essays on Photography,* ed. Alan Trachtenberg. New Haven, Conn.: Leete's Island, 1980.

Bennett, W. Lance. *News: The Politics of Illusion.* New York: Longman, 1988.

Berger, Jean, and Jean Mohr. *Another Way of Telling.* New York: Pantheon, 1982.

Boorstin, Daniel J. *The Image: A Guide to Pseudo-Events in America.* New York: Atheneum, 1978.

Boulding, Kenneth E. *The Image.* Ann Arbor: University of Michigan Press, 1956.

Brooks, Courtney G., James M. Grimwood, and Loyd S. Swenson, Jr. *Chariots for Apollo: A History of Manned Lunar Spacecraft.* Washington, D.C.: NASA, 1979.

Brown, Seyon. "Perceived Deficiencies in the Nation's Power." In *John F. Kennedy and Presidential Power,* ed. Earl Latham. Lexington: Heath, 1972.

Burke, Kenneth. *Counter-Statement.* Berkeley: University of California Press, 1968.

Carpenter, Ronald. *The Eloquence of Frederick Jackson Turner.* San Marino, Calif.: Huntington Library, 1983.

Cirino, Robert. "To the Moon: 'There Really Isn't Any Argument.'" In *Don't Blame the People*. Los Angeles: Diversity, 1971.

Davidson, Dorothy P., ed. *Book Review Digest*. New York: Wilson, 1963.

Denton, Robert E., Jr., and Dan F. Hahn. *Presidential Communication: Description and Analysis*. New York: Praeger, 1986.

Diamond, Edwin. *The Rise and Fall of the Space Age*. Garden City, N.Y.: Doubleday, 1964.

Edelman, Murray. *Political Language: Words That Succeed and Policies That Fail*. New York: Academic Press, 1977.

———. *Politics as Symbolic Action: Mass Arousal and Quiescence*. Chicago: Markham, 1971.

Edey, Maitland. *Great Photographic Essays from "Life."* New York: Little, Brown, 1978.

Edwards, George C. *The Public Presidency*. New York: St. Martin's, 1983.

Eisenhower, Dwight D. *Waging Peace*. New York: Doubleday, 1965.

Elson, Robert T. *The World of Time, Inc.: The Intimate History of a Publishing Enterprise*. ed. Duncan Norton-Taylor. Vol. 1. New York: Atheneum, 1973.

Emme, Eugene M. *A History of Space Flight*. New York: Holt, 1965.

Etzioni, Amitai. *The Moon-Doggle: Domestic and International Implications of the Space Race*. Garden City, N.Y.: Doubleday, 1964.

Fairlie, Henry. *The Kennedy Promise: The Politics of Expectation*. Garden City, N.Y.: Doubleday, 1973.

Fallaci, Oriana. *If the Sun Dies*. New York: Atheneum, 1966.

Fisher, Walter R. *Human Communication as Narration: Toward a Philosophy of Reason, Value, and Action*. Columbia: University of South Carolina Press, 1987.

Gallup, George. *The Gallup Poll: Public Opinion, 1935–1971*. Vol. 3. New York: Random House, 1972.

Gans, Herbert J. *Deciding What's News*. New York: Vintage, 1979.

Gibney, Frank B., and George Feldman, Jr. *The Reluctant Space-Farers: The Political and Economic Consequences of America's Space Effort*. New York: New American Library, 1965.

Gilder, George. *Wealth and Poverty*. New York: Basic, 1981.

Griffith, Thomas. *How True: A Skeptic's Guide to Believing the News*. Toronto: Little, Brown, 1974.

Grissom, Betty, and Henry Still. *Starfall*. New York: Crowell, 1974.

Halberstam, David. *The Powers That Be*. New York: Norton, 1977.

Hamblin, Dora Jane. *That Was the Life*. New York: Norton, 1977.

Hart, Roderick P. *Verbal Style and the Presidency: A Computer-Based Analysis*. Orlando: Academic Press, 1984.

Head, Sidney W. *Broadcasting in America*. Boston: Houghton Mifflin, 1972.

Hechler, Ken. *The Endless Space Frontier: A History of the House Committee on Science and Astronautics, 1959–1978*. San Diego: American Astronautical Society, 1982.

———. *Toward the Endless Frontier: A History of the Committee on Science and Technology, 1959–1979*. Washington, D.C.: GPO, 1980.

Hicks, Wilson. *Words and Pictures*. New York: Arno, 1973.

Hogan, J. Michael. *The Panama Canal in American Politics: Domestic Advocacy and the Evolution of Policy.* Carbondale: Southern Illinois University Press, 1986.

Holmes, Jay. *America on the Moon: The Enterprise of the Sixties.* Philadelphia: Lippincott, 1962.

The Ideas of Henry Luce. Ed. John K. Jessup. New York: Atheneum, 1969.

Jamieson, Kathleen Hall, and Karlyn Kohrs Campbell. *The Interplay of Influence.* Belmont, Calif.: Wadsworth, 1983.

Johnson, Lyndon Baines. *The Vantage Point: Perspectives of the Presidency, 1963–1969.* New York: Holt, 1971.

Jussim, Estelle. *Visual Communication and the Graphic Arts.* New York: Bowker, 1974.

Keefe, William J., and Morris S. Ogul. *The American Legislative Process: Congress and the States.* Englewood Cliffs, N.J.: Prentice-Hall, 1993.

Killiam, James R., Jr. *Sputnik, Scientists, and Eisenhower.* Cambridge: MIT Press, 1977.

Levine, Arthur L. *The Future of the U.S. Space Program.* New York: Praeger, 1975.

Lewis, Richard L. *Appointment on the Moon.* New York: Viking, 1968.

———. *The Voyages of Apollo: The Exploration of the Moon.* New York: Quadrangle, 1974.

Logsdon, John M. *The Decision to Go to the Moon: Project Apollo and the National Interest.* Cambridge: MIT Press, 1970.

Lucas, Stephen E. *Portents of Rebellion: Rhetoric and Revolution in Philadelphia, 1765–1776.* Philadelphia: Temple University Press, 1976.

McDougall, Walter A. *The Heavens and the Earth: A Political History of the Space Age.* New York: Basic, 1985.

MacIntyre, Alasdair. *After Virtue: A Study in Moral Theory.* Notre Dame, Ind.: University of Notre Dame Press, 1981.

Mailer, Norman. *Of a Fire on the Moon.* Boston: Little, Brown, 1969.

Martin, Ralph G. *A Hero for Our Time: An Intimate Story of the Kennedy Years.* New York: Macmillan, 1983.

Mazlish, Bruce, ed. *The Railroad and the Space Program: An Exploration in Historical Analogy.* Cambridge: MIT Press, 1965.

Murray, Charles, and Catherine Bly Cox. *Apollo: The Race to the Moon.* New York: Simon and Schuster, 1989.

Neustadt, Richard E. *Presidential Power: The Politics of Leadership.* New York: Wiley, 1960.

Nieburg, H. L. *In the Name of Science.* Chicago, Quadrangle, 1966.

Nimmo, Dan, and James E. Coombs. *Mediated Political Realities.* New York: Longman, 1983.

Nimmo, Dan, and Robert L. Savage. *Candidates and Their Images.* Pacific Palisades, Calif.: Goodyear, 1976.

Paper, Lewis J. *The Promise and the Performance: The Leadership of John F. Kennedy.* New York: Crown, 1975.

Patterson, Thomas E. *The Mass Media Election: How Americans Choose Their President.* New York: Praeger, 1980.

Postman, Neil. *Amusing Ourselves to Death.* New York: Penguin, 1985.

Prendergast, Curtis, and Geoffrey Colvin. *The World of Time, Inc.: The Intimate History of a Publishing Enterprise*. Ed. Robert Lubar. Vol. 3. New York: Atheneum, 1986.

Rhode, Robert, and Floyd McCall. *Press Photography: Reporting with a Camera*. New York: Macmillan, 1961.

Schlesinger, Arthur M., Jr. *A Thousand Days: John F. Kennedy in the White House*. Boston: Houghton Mifflin, 1965.

Shaw, Donald L., and Maxwell E. McCombs. *The Emergence of American Political Issues: The Agenda-Setting Function of the Press*. St. Paul, Minn.: West Publishing, 1977.

Sidey, Hugh. *John F. Kennedy, President*. New York: Atheneum, 1964.

Slotkin, Richard. *The Fatal Environment: The Myth of the Frontier in the Age of Industrialization, 1800–1890*. New York: Atheneum, 1985.

Smith, Henry Nash. *Virgin Land: The American West as Symbol and Myth*. Cambridge: Harvard University Press, 1978.

Smith, Steven S., and Christopher J. Deering. *Committees in Congress*. Washington, D.C.: Congressional Quarterly Press, 1984.

Sontag, Susan. *On Photography*. New York: Farrar, Straus, and Giroux, 1977.

Sorensen, Theodore C. *Kennedy*. New York: Harper, 1965.

Sullivan, Walter, ed. *America's Race for the Moon*. New York: Random House, 1962.

Swanberg, W. A. *Luce and His Empire*. New York: Scribner's, 1972.

Swenson, Loyd S., Jr., James M. Grimwood, and Charles C. Alexander. *This New Ocean: A History of Project Mercury*. Washington, D.C.: NASA, 1966.

Van Dyke, Vernon. *Pride and Power: The Rationale of the Space Program*. Urbana: University of Illinois Press, 1964.

Van Riper, Frank. *Glenn: The Astronaut Who Would Be President*. New York: Empire, 1983.

Wainwright, Loudon. *The Astronauts: Pioneers in Space*. New York: Golden Press, 1961.

———. *The Great American Magazine: An Inside History of "Life."* New York: Knopf, 1986.

Wattenberg, Ben J. *The Statistical History of the United States*. New York: Basic, 1976.

We Seven. New York: Simon and Schuster, 1962.

Wilson, Woodrow. *Congressional Government*. Baltimore: Johns Hopkins University Press, 1981.

Windt, Theodore. *Presidential Rhetoric, 1961–1980*. Dubuque: Kendall/Hunt, 1980.

———, ed. *Essays in Presidential Rhetoric*. Dubuque: Kendall/Hunt, 1983.

Wolfe, Tom. *The Right Stuff*. New York: Bantam, 1980.

Wood, James P. *Magazines in the United States*. New York: Ronald Press, 1971.

The World Almanac and Book of Facts for 1962. New York: New York World-Telegraph, 1963.

The World Almanac and Book of Facts for 1963. New York: New York World-Telegraph, 1964.

Young, Hugo, Brian Silcock, and Peter Dunn. *Journey to Tranquility.* Garden City, N.Y.: Doubleday, 1970.
Zarefsky, David. *President Johnson's War on Poverty: Rhetoric and History.* University: University of Alabama Press, 1986.

Articles

Adams, William C., Allison Salzman, William Vantine, Leslie Suelter, Anne Baxter, Lucille Bonvouloir, Barbara Brenner, Margaret Ely, Jean Feldman, and Ron Ziegel. "The Power of *The Right Stuff:* A Quasi-Experimental Field Test of the Docudrama Hypothesis." *Public Opinion Quarterly* 49 (1985): 330–39.
Almond, Gabriel A. "Public Opinion and the Development of Space Technology." *Public Opinion Quarterly* 24 (1960): 553–72.
Barger, Harold M. "Images of Political Authority in Four Types of Black Newspaper." *Journalism Quarterly* 50 (1973): 645–51.
Barrett, Harold. "John F. Kennedy Before the Greater House Ministerial Association." *Central States Speech Journal* 15 (1964): 259–66.
Berg, David M. "Rhetoric, Reality, and Mass Media." *Quarterly Journal of Speech* 58 (1972): 255–63.
Bernstein, Norbert. "Book Reviews." *Library Journal* 1 November 1962: 4034.
Berthold, Carol A. "Kenneth Burke's Cluster-Agon Method: Its Development and an Application." *Central States Speech Journal* 27 (1976): 302–09.
Bitzer, Lloyd F. "Rhetorical Public Communication." *Critical Studies in Mass Communication* 4 (1987): 425–27.
Boot, William. "NASA and the Spellbound Press." *Columbia Journalism Review* 25.4 (1986): 23–29.
Boylan, James. "Declaration of Independence." *Columbia Journalism Review* 25.5 (1986): 29–46.
Carpenter, Ronald. "Frederick Jackson Turner and the Rhetorical Impact of the Frontier Thesis." *Quarterly Journal of Speech* 63 (1977): 117–29.
"The Case for Space." *New Republic* 27 June 1988: 7–9.
Ceasar, James W., Glen E. Thurow, Jeffrey Tulis, and Joseph M. Bessette. "The Rise of the Rhetorical Presidency." *Presidential Studies Quarterly* 10 (1980): 158–71.
Cowen, R. C. "Ordinary Supermen." *Christian Science Monitor* 15 November 1962: 14.
Depoe, Stephen P. "Space and the 1960 Presidential Campaign: Kennedy, Nixon, and 'Public Time.'" *Western Journal of Speech Communication* 55 (Spring 1991): 215–33.
Etzioni, Amitai. "When Scientists Testify." *Bulletin of the Atomic Scientists* 20 October 1964: 23–26.
"Expanded Space Effort Viewed as Cold War Need, Poll Shows." *Aviation Week and Space Technology* 5 February 1962: 34.

Farrell, Thomas B. "Knowledge, Consensus, and Rhetorical Theory." *Quarterly Journal of Speech* 62 (1976): 1–14.

Farrell, Thomas B., and Thomas G. Goodnight. "Accidental Rhetoric: The Root Metaphors of Three Mile Island." *Communication Monographs* 48 (1981): 271–300.

Fisher, Walter. "Narration as a Human Communication Paradigm: The Case of Public Moral Argument." *Communication Monographs* 51 (1984): 1–22.

———. "Rhetorical Fiction and the Presidency." *Quarterly Journal of Speech* 66 (1980): 119–26.

Fitzpatrick, John R. "Congressional Debating." *Quarterly Journal of Speech* 27 (1941): 251–55.

Folkerts, Jean Lange. "William Allen White's Anti-Populist Rhetoric as an Agenda-Setting Technique." *Journalism Quarterly* 60 (1983): 28–34.

Gans, Herbert J. "The Messages Behind the News." *Columbia Journalism Review* 17.1 (1979): 40–45.

Golden, James L. "John F. Kennedy and the 'Ghosts.'" *Quarterly Journal of Speech* 52 (1966): 348–57.

Goodnight, Thomas G. "The Personal, Technical, and Public Spheres of Argument: A Speculative Inquiry into the Art of Public Deliberation." *Journal of the American Forensic Association* 18 (1982): 214–27.

Hahn, Dan F. "Ask Not What a Youngster Can Do for You: Kennedy's Inaugural Address." *Presidential Studies Quarterly* 12 (1982): 610–14.

Hankins, Sara R. "Archetypal Alloy: Reagan's Rhetorical Image." *Central States Speech Journal* 33 (1983): 33–43.

Harding, H. F. "John F. Kennedy: Campaigner." *Quarterly Journal of Speech* 46 (1960): 362–64.

Hart, John. "Assessing Presidential Leadership: A Comment on Williams and Kershaw." *Political Studies* 28 (1980): 567–78.

Hightower, Paul. "Readers See What They Want to See." *Grassroots Editor.* (June-July 1976): 13–14.

———. "A Study of the Messages in Depression-Era Photos." *Journalism Quarterly* 57 (Autumn 1980): 495–497.

Hill, David B. "Viewer Characteristics and Agenda Setting by Television News." *Public Opinion Quarterly* 49 (1985): 340–50.

Hofsetter, Richard C., Cliff Zunkin, and Terry F. Buss. "Political Imagery and Information in an Age of Television." *Journalism Quarterly* 55 (1978): 562–69.

Hutchison, Earl R. "Kennedy and the Press: The First Six Months." *Journalism Quarterly* 38 (1961): 453–59.

Jamieson, Kathleen, and Karlyn Kohrs Campbell. "Rhetorical Hybrids: Fusions of Generic Elements." *Quarterly Journal of Speech* 68 (1982): 146–57.

Jaroff, Leon. "Onward to Mars." *Time* 18 July 1988: 46–53.

Kane, Peter E. "Evaluating the 'Great Debates.'" *Western Journal of Speech Communication* 30 (Spring 1960): 89–96.

Kenny, Edward B. "Another Look at Kennedy's Inaugural Address." *Communication Quarterly* 13 (November 1965): 17–19.

Kerr, Harry P. "John F. Kennedy." *Quarterly Journal of Speech* 46 (1960): 241.

———. "The President and the Press." *Western Journal of Speech Communication* 27 (1983): 216–21.

Lemert, Timothy M., and Kenneth G. Sheinkopf. "Television News and Status Conferral." *Journal of Broadcasting* 14 (1970): 491–97.

Lewis, William F. "Telling America's Story: Narrative Form and the Reagan Presidency." *Quarterly Journal of Speech* 73 (1987): 280–302.

McCombs, Maxwell E., and Donald L. Shaw. "The Agenda-Setting Function of the Mass Media." *Public Opinion Quarterly* 36 (1972): 176–87.

Mechling, Elizabeth W. "Patricia Hearst: MYTH AMERICA 1974, 1975, 1976." *Western Journal of Speech Communication* 43 (1979): 168–79.

Meyers, Renee A., Thomas L. Newhouse, and Dennis E. Garrett. "Political Momentum: Television News Treatment." *Communication Monographs* 45 (1978): 382–88.

Michael, Donald N. "The Beginning of the Space Age and American Public Opinion." *Public Opinion Quarterly* 24 (1960): 573–82.

O'Keefe, Timothy M., and Kenneth G. Skeinkopf. "The Voter Decides: Candidate Image or Campaign Issue?" *Journal of Broadcasting* 18 (1974): 403–12.

Osborne, Leonard L. "Rhetorical Patterns in President Kennedy's Major Speeches: A Case Study." *Presidential Studies Quarterly* 10 (1980): 332–35.

Ostman, Ronald E., and William A. Babcock. "Three Major Newspapers' Content and President Kennedy's Press Conference Statements Regarding Space Exploration and Technology." *Presidential Studies Quarterly* 13 (1983): 111–20.

"Packaging Bravery." *New Republic* 26 March 1962: 2.

"Parables of the Space Age—the Ideological Basis of Space Exploration." *Western Folklore* 46 (October 1987): 227–93.

Payne, Seth, and William D. Marbach. "Science Is Still Waiting for Liftoff." *Business Week* 17 October 1988: 50–51.

Philport, Joseph C., and Robert E. Balon. "Candidate Image in a Broadcast Debate." *Journal of Broadcasting* 19 (1975): 181–93.

Pollard, James E. "The Kennedy Administration and the Press." *Journalism Quarterly* 41 (1964): 3–14.

"The Polls: Defense, Peace, and Space." *Public Opinion Quarterly* 25 (1961): 478–89.

Pratt, James W. "An Analysis of the Three Crisis Speeches." *Western Journal of Speech Communication* 34 (1970): 194–203.

Rowland, Robert C. "The Relationship between the Public and the Technical Spheres of Argument: A Case Study of the Challenger Disaster." *Central States Speech Journal* 37 (1986): 136–46.

Rushing, Janice Hocker. "Mythic Evolution of 'The New Frontier' in Mass Mediated Rhetoric." *Critical Studies in Mass Communication* 3 (1986): 265–96.

———. "The Rhetoric of the American Western Myth." *Communication Monographs* 50 (1983): 14–32.

———. "Ronald Reagan's 'Star Wars' Address: Mythic Containment of Technical Reasoning." *Quarterly Journal of Speech* 72 (1986): 415–33.
Samovar, Larry A. "Ambiguity and Unequivocation in the Kennedy-Nixon Television Debates." *Quarterly Journal of Speech* 48 (1962): 277–79.
———. "Ambiguity and Unequivocation in the Kennedy-Nixon Television Debates: A Rhetorical Analysis." *Western Journal of Speech Communication* 29 (1965): 211–18.
Sherrod, Robert. "The Selling of the Astronauts." *Columbia Journalism Review* 12.3 (1973): 16–25.
Sojka, G. S. "The Astronaut: An American Hero with 'The Right Stuff.'" *Journal of American Culture* 7.3 (1984): 118–21.
Stelzner, Hermann G. "Humphrey and Kennedy Court West Virginia, May 3, 1960." *Southern Speech Communication Journal* 37 (Fall 1971): 21–33.
Stewart, Charles J. "The Pulpit in Time of Crisis: 1865 and 1963." *Communication Monographs* 32 (1965): 427–34.
Talmer, Jerry. "Rocket to the Moon." *Village Voice* 1 March 1962: 4.
Tsang, Kuo-jen. "News Photos in Time and Newsweek." *Journalism Quarterly* 61 (Autumn 1984): 578–84.
Wellborn, Stanley. "Man's Inevitable Trip to Mars." *U.S. News and World Report* 3 August 1987: 46–48.
Williams, Robert J. "Apologists and Revisionists: A Rejoinder to Hart." *Political Studies* 28 (1980): 579–83.
Williams, Robert J., and David A. Kershaw. "Kennedy and Congress: The Struggle for the New Frontier." *Political Studies* 27 (1979): 390–404.
Windt, Theodore O., Jr. "Presidential Rhetoric: Definition of a Field of Study." *Central States Speech Journal* 35 (1984): 24–34.
Wolfarth, David L. "John F. Kennedy in the Tradition of Inaugural Speeches." *Quarterly Journal of Speech* 47 (1961): 124–32.
Woodburn, B. W. "Reader Interest in Newspaper Pictures." *Journalism Quarterly* 24 (September 1947): 197–201.

Newspapers and Magazines

Life. 1 January 1959 to 31 December 1963.
Look. 1 January 1959 to 31 December 1963.
Newsweek. 1 January 1959 to 31 December 1963.
New York Times. 1 January 1959 to 31 December 1963.
Popular Mechanics. 1 January 1959 to 31 December 1963.
Popular Science. 1 January 1959 to 31 December 1963.
Reader's Digest. 1 January 1959 to 31 December 1963.
Saturday Evening Post. 1 January 1959 to 31 December 1963.
Time. 1 January 1959 to 31 December 1963.

Public Documents

Bush, George. *Weekly Compilation of Presidential Documents.* Washington, D.C.: GPO, 1989.

Eisenhower, Dwight D. *Public Papers of the Presidents: Dwight D. Eisenhower, 1958, 1959, 1960.* Washington, D.C.: GPO, 1959–61.

Kennedy, John F. *Public Papers of the Presidents: John F. Kennedy, 1961, 1962, 1963.* Washington, D.C.: GPO, 1962–64.

National Aeronautics and Space Administration. *Conference on Space-Age Planning.* Washington, D.C.: GPO, 1963.

———. *Impact of Progress in Space on Science.* Washington, D.C.: GPO, 1962.

———. *Manned Space Flight—1963.* Washington, D.C.: GPO, 1963.

———. *One-Two-Three and the Moon: Projects Mercury, Gemini, and Apollo of America's Manned Space Flight Program.* Washington, D.C.: GPO, 1963.

———. *Proceedings of the First National Conference on the Peaceful Uses of Space.* Washington, D.C.: GPO, 1961.

———. *Proceedings of the Second National Conference on the Peaceful Uses of Space.* Washington, D.C.: GPO, 1963.

———. *Space, the New Frontier.* Washington, D.C.: GPO, 1962.

———. *Space, the New Frontier.* Washington, D.C.: GPO, 1963.

United States. Cong. *Congressional Record.* Vols. 107 (1961), 108 (1962), 109 (1963).

United States. Cong. House. Committee on Appropriations. *Independent Offices Appropriations for 1962.* 87th Cong., 1st sess. Washington, D.C.: GPO, 1961.

———. *Independent Offices Appropriations for 1963.* 87th Cong., 2d sess. Washington, D.C.: GPO, 1962.

———. *Second Supplemental Appropriation Bill, 1962.* 87th Cong., 2d sess. Washington, D.C.: GPO, 1961.

United States. Cong. House. Committee on Science and Astronautics. *Discussion of Soviet-in-Space Shot.* 87th Cong., 1st sess. Washington, D.C.: GPO, 1961.

———. *1962 NASA Authorization, Hearings on H.R. 6874.* 87th Cong., 1st sess. Washington, D.C.: GPO, 1961.

———. *1963 NASA Authorization, Hearings on H.R. 11737.* 87th Cong., 2d sess. Washington, D.C.: GPO, 1962.

———. *1964 NASA Authorization, Hearings on H.R. 7500.* 88th Cong., 1st sess. Washington, D.C.: GPO, 1963.

———. *The Practical Values of Space Exploration.* 87th Cong., 1st sess. Washington, D.C.: GPO, 1961.

United States. Cong. House. Committee of the Whole House on the State of the Union. *Authorizing Appropriations to the National Aeronautics and Space Administration: Report to Accompany H.R. 6874.* 87th Cong., 1st sess. H. Rept. 391. Washington, D.C.: GPO, 1961.

———. *Authorizing Appropriations to the National Aeronautics and Space Administration: Report to Accompany H.R. 11737.* 87th Cong., 2d sess. H. Rept. 1674. Washington, D.C.: GPO, 1962.

———. *Authorizing Appropriations to the National Aeronautics and Space Administration: Report to Accompany H.R. 7500.* 88th Cong., 2d sess. H. Rept. 591. Washington, D.C.: GPO, 1963.

United States. Cong. House. Select Committee on Astronautics and Space Sciences. *Astronautics and Space Exploration: Hearings on H.R. 11881.* 85th Cong., 2d sess. H. Rept. 591. Washington, D.C.: GPO, 1958.

———. *Independent Offices Appropriations, 1963.* 87th Cong., 2d sess. Washington, D.C.: GPO, 1962.

———. *Independent Offices Appropriations, 1964.* 88th Cong., 1st sess. Washington, D.C.: GPO, 1963.

———. *Supplemental Appropriation Bill for 1962.* 87th Cong., 1st sess. Washington, D.C.: GPO, 1961.

United States. Cong. Senate. Committee of Aeronautical and Space Sciences. *Authorizing Appropriations to the National Aeronautics and Space Administration: Report to Accompany H.R. 6874.* 87th Cong., 1st sess. S. Rept. 475. Washington, D.C.: GPO, 1961.

———. *Authorizing Appropriations to the National Aeronautics and Space Administration: Report to Accompany H.R. 11737.* 87th Cong., 2d sess. S. Rept. 1633. Washington, D.C.: GPO, 1962.

———. *Authorizing Appropriations to the National Aeronautics and Space Administration: Report to Accompany H.R. 7500.* 88th Cong., 1st sess. S. Rept. 385. Washington, D.C.: GPO, 1963.

———. *NASA Authorization for Fiscal Year 1962, Hearings on H.R. 6874.* 87th Cong., 1st sess. Washington, D.C.: GPO, 1961.

———. *NASA Authorization for Fiscal Year 1963, Hearings on H.R. 11737.* 87th Cong., 2d sess. Washington, D.C.: GPO, 1962.

———. *NASA Authorization for Fiscal Year 1964, Hearings on S. 1245.* 88th Cong., 1st sess. Washington, D.C.: GPO, 1963.

———. *Orbital Flight of John H. Glenn, Jr.* 87th Cong., 1st sess. Washington, D.C.: GPO, 1962.

———. *Scientists' Testimony on Space Goals.* 88th Cong., 1st sess. Washington, D.C.: GPO, 1963.

United States Cong. Senate. Committee on Appropriations. *Independent Offices Appropriations, 1962.* 87th Cong., 1st sess. Washington, D.C.: GPO, 1961.

Unpublished Documents

"Astronauts Press Conference." 16 September 1959. NASA History Office, Washington, D.C.

Bonny, Walter. Memorandum to Herb Rosen. "Name of NASA Vehicles." 12 January 1960. NASA History Office, Washington, D.C.

Duff, Brian. Memorandum to All Public Information Officers. 20 June 1963. NASA History Office, Washington, D.C.

Jahnige, Thomas P. "Congress and Space." Ph.D. diss., Claremont Graduate School, 1965.

Johnson, John. Address. "The New Frontiers of Space." 24 March 1961. NASA History Office, Washington, D.C.

Johnson, Lyndon B. Address. "American Institute of Aeronautics and Astro-

nautics Second Manned Space Flight Meeting." 23 April 1963. NASA History Office, Washington, D.C.
————. Address. "Goddard Memorial Award Dinner." 14 November 1963. NASA History Office, Washington, D.C.
————. Address. "National Rocket Club Dinner." 23 March 1963. NASA History Office, Washington, D.C.
————. Address. "Remarks at Space Center Dedication." 14 November 1963. NASA History Office, Washington, D.C.
————. "Memorandum for the President: Evaluation of the Space Program." 28 April 1961. NASA History Office, Washington, D.C.
Kennedy, John F. "Memorandum for the Vice President." 20 April 1961. NASA History Office, Washington, D.C.
Kerr, James R. "Congressmen as Overseers: Surveillance of the Space Program." Ph.D. diss., Stanford University, 1963.
Lingle, Walter L., Jr. Memorandum to Hugh Dryden. "Policy Concerning the Astronauts." 1 August 1962. NASA History Office, Washington, D.C.
Lloyd, Bill. Memorandum to James Webb. "Anticipated Questions from News Media." 21 May 1963. NASA History Office, Washington, D.C.
"MA-7 Press Conference." Transcript. 27 May 1962. NASA History Office, Washington, D.C.
"Mercury-Atlas 9." NASA News Release. 10 May 1963. NASA History Office, Washington, D.C.
"Mercury-Redstone 3." NASA News Release. 26 April 1961. NASA History Office, Washington, D.C.
"MR-4 Design Changes." NASA News Release. 16 July 1961. NASA History Office, Washington, D.C.
"National Aeronautics and Space Administration News Conference." Transcript. 22 July 1961. NASA History Office, Washington, D.C.
"Presentation of NASA Distinguished Service Awards." Transcript. 24 May 1962. NASA History Office, Washington, D.C.
"Press Conference: Astronaut Program Outlined." Transcript. 12 May 1959. NASA History Office, Washington, D.C.
Sheer, Julian. "NASA News Release." 1 May 1961. NASA History Office, Washington, D.C.
Shepard, Alan. Letter to P. Michael Whye. 11 November 1976. NASA History Office, Washington, D.C.
Thompson, Edward. Letter to James Webb. 30 March 1963. NASA History Office, Washington, D.C.
Webb, James. Address. "Remarks to the American Association of School Administrators." 18 February 1962. NASA History Office, Washington, D.C.
————. "Administrator's Presentation to the President." 28 April 1961. NASA History Office, Washington, D.C.
————. Letter to Edward Thompson. 23 April 1962. NASA History Office, Washington, D.C.
————. "NASA News Release." 1 May 1961. NASA History Office, Washington, D.C.

Index

pioneers, 36, 43, 82; situational constraints and, 39; and technology, 136. *See also* Narratives; Project Apollo

Fulbright, William, 119, 120, 121, 122, 124, 126, 131

Fulton, James, 95, 96, 97, 99, 100, 101, 102, 104, 121, 124, 126

Fulton, Richard, 129

Gagarin, Yuri, 3, 6, 20, 35, 40, 62, 79, 81, 95, 97, 101, 102, 111, 114, 118

Gans, Herbert, 53, 54, 56, 63, 64

Glenn, John H., Jr., 1, 6, 7, 15, 16, 21, 32, 34, 35, 37, 38, 39, 41, 42–45, 46, 47, 48, 57, 60–61, 62, 63, 64–65, 69, 74, 75, 80, 81–83, 85, 86, 87, 88, 89, 90, 102, 103, 128–29, 130, 133

Goldwater, Barry, 123

Grissom, Gus, 16, 34, 38, 41, 42, 43, 45, 48, 57, 81, 85, 86, 87, 90

Gross, H. R., 117, 118, 120

Gruening, Ernest, 119

Gurney, Edward, 113

Halberstam, David, 68, 70

Hamblin, Dora Jane, 69, 70, 72, 82

Hechler, Ken, 110, 125

Hicks, Wilson, 75

Hightower, Paul, 76

Holmes, D. Brainerd, 24, 44, 45

Holland, Spessard, 97, 103

House Committee on Science and Astronautics, 99; hearings, 94; and NASA budgets, 94; reaction to Gagarin flight, 95, 101, 118; as advocates for space program, 96; and criticism of NASA budgets, 97; and moon race, 97; and frontier narrative, 98; and military value of Project Apollo, 98–101; and pride and space race, 101–2; misleading committee reports of, 111–15; and criticism of Project Apollo, 113–14

Kennedy, Edward, 56

Kennedy, John F., 1–13, 15, 17–20, 21, 22, 23, 24, 25, 26, 27, 28, 29, 30, 31–33, 34, 35, 36, 41, 43, 73, 90, 95, 121, 130, 132, 137; and NASA budgets, 2, 3, 5, 17, 31, 36; and New Frontier, 2, 3, 5, 6, 29, 31, 32, 37, 133, 137; and space policy, 2, 3, 12; and

25 May address, 3, 13, 20, 31; and technocracy, 4; and campaign of 1960, 6; and Project Apollo, 13, 20–28; and Rice University address, 15, 33; and joint moon program, 18, 95, 121

Kerr, James, 117

Kerr, Robert, 56, 102, 110, 125, 126, 129

King, David, 97, 101, 120

Kush, Polykarp, 106, 108, 122, 124

Lane, Thomas, 128, 130

LeMay, Curtis, 36, 122

Lewis, William, 90, 131

Life, 68–92; and astronauts' contract, 18, 55, 71–75, 92, 133; photographs in, 70, 88; image-making and astronauts, 80–82, 84–85, 91; and *The Astronauts*, 84; and *We Seven*, 84; and astronauts' wives, 86–87; photographic narratives in, 88–90

Lindsay, John, 130, 131

Lloyd, Bill, 7, 22

Logsdon, John, 2, 7

Low, George, 21, 23, 33, 38, 39

Luce, Henry Robinson, 68, 69, 70, 71, 72, 84, 92

McDougall, Walter, 1, 2, 3, 4, 6, 7, 30, 45, 136

Magnuson, Warren, 98, 102, 103

Mansfield, Mike, 129

Martin, Ralph, 3, 31

Mass media, 9, 40, 50–68, 133, 135; and enduring values, 53–67; and negative coverage of space program, 53–56; and small-town pastoralism, 53–54; and support of manned space program, 53, 58; and altruistic democracy, 54–56; and individualism, 56–67; and positive coverage of space program, 56–67; and frontier narrative, 59–61

Merrow, Chester, 128

Miller, George, 16, 17, 19, 97, 101, 110, 111, 112, 113, 117, 118, 121, 123, 124

Mohr, Jean, 76

Morse, Ralph, 88, 89, 92

Mundt, Karl, 128

Narratives, 4, 5, 30, 31, 132, 133

National Aeronautics and Space Administration (NASA): budgets, 3, 9, 13, 17, 19, 94, 95, 104, 117, 118; and